GOLD METAL WATERS

GOLD METAL WATERS

THE ANIMAS RIVER AND THE GOLD KING MINE SPILL

EDITED BY

Brad T. Clark and Pete McCormick

UNIVERSITY PRESS OF COLORADO
Louisville

© 2021 by University Press of Colorado

Published by University Press of Colorado
245 Century Circle, Suite 202
Louisville, Colorado 80027

All rights reserved
First paperback edition 2022
Manufactured in the United States of America

The University Press of Colorado is a proud member of the Association of University Presses.

The University Press of Colorado is a cooperative publishing enterprise supported, in part, by Adams State University, Colorado State University, Fort Lewis College, Metropolitan State University of Denver, Regis University, University of Colorado, University of Northern Colorado, University of Wyoming, Utah State University, and Western Colorado University.

∞ This paper meets the requirements of the ANSI/NISO Z39.48–1992 (Permanence of Paper).

ISBN: 978-1-64642-174-9 (hardcover)
ISBN: 978-1-64642-308-8 (paperback)
ISBN: 978-1-64642-175-6 (ebook)
https://doi.org/10.5876/9781646421756

Library of Congress Cataloging-in-Publication Data

Names: Clark, Brad T., editor. | McCormick, Pete, 1971– editor.
Title: Gold metal waters : the Animas River and the Gold King Mine spill / edited by Brad T. Clark and Pete McCormick.
Description: Louisville : University Press of Colorado, [2021] | Includes bibliographical references and index.
Identifiers: LCCN 2021001160 (print) | LCCN 2021001161 (ebook) | ISBN 9781646421749 (hardcover) | ISBN 9781646423088 (paperback) | ISBN 9781646421756 (ebook)
Subjects: LCSH: Acid mine drainage—Colorado—Gold King Mine (San Juan County) | Abandoned mined lands reclamation—Accidents—Colorado—Gold King Mine (San Juan County) | Waste spills—Colorado—Gold King Mine (San Juan County) | Water—Pollution—Animas River (Colo. and N.M.) | Hard rock mines and mining—Environmental aspects—West (U.S.)
Classification: LCC TD427.A28 G65 2021 (print) | LCC TD427.A28 (ebook) | DDC 363.17/90978829—dc23
LC record available at https://lccn.loc.gov/2021001160
LC ebook record available at https://lccn.loc.gov/2021001161

Front-cover photograph: Jerry McBride/*Durango Herald*/Polaris. Back-cover illustration: courtesy, Center of Southwest Studies, Fort Lewis College, Durango, Colorado.

For the people of the Animas Basin who have shown remarkable resilience, time and time again.

Contents

GOLD METAL **WATERS**

INTRODUCTION

From Gold *Medal* to Gold *Metal* Waters

BRAD T. CLARK

More than 500,000 abandoned hardrock mines are scattered across the American West, a legacy of the boom-and-bust cycle of resource development. Estimates for comprehensive cleanup range from $36 million to $72 billion (Moyer 2016). At many mine sites, acidic mixes of heavy metals have drained unchecked for decades from the myriad shafts, tunnels, and portals (or so-called adits).[1] Degraded water quality and damage to aquatic environments have resulted across many regional watersheds. According to the US Environmental Protection Agency (EPA), abandoned hardrock mines affect 40 percent of headwaters in the western United States; an additional 180,000 acres of lakes and reservoirs are estimated to have been impacted (Limerick et al. 2005). Since many of these affected waters are sourced from or flow across public lands, the lost revenue for communities with economies heavily dependent on any array of outdoor activities (e.g., angling, hiking, boating) is substantial.[2]

From a national perspective, millions of dollars are lost each year from the paucity of royalties paid by private enterprises on the wealth they've

DOI: 10.5876/9781646421756.c000

extracted from beneath the public domain. Ever since manifest destiny lured explorers and fortune seekers west, profits from hardrock mining have been privatized while environmental impacts remained socialized—culminating in what Pulitzer Prize–winning historian Vernon L. Parrington (1930) referred to as "the Great Barbecue" of the American West.

In Colorado alone, the Colorado Division of Reclamation, Mining, and Safety (CDRMS) estimates that there are *at least* 23,000 abandoned hardrock mine lands—classified as lands where mines operated prior to 1975, when the state began to establish limited forms of mining and reclamation standards. Today, these hardrock mine lands impact water quality in approximately 1,645 miles of streams and rivers.[3]

Remediation efforts have been mixed, often stymied by a combination of outdated laws, funding woes, and ill-enforced regulations. Local politics, persistent NIMBY-ism, and liability concerns have further frustrated policy development and comprehensive restoration efforts—including National Priorities List (NPL) designation under the 1980 Comprehensive Environmental Response, Compensation, and Liability Act (CERCLA), henceforth referred to as Superfund. All the while, acid mine drainage (AMD) from many of these "ticking time bombs" has contaminated watersheds and river basins on which tens of millions of westerners increasingly depend.

> Whether it's 100,000 or 500,000 [abandoned mines], that's hundreds of thousands too many . . . the Animas River spill has alerted the nation to the much more broad problem that many people were not paying attention to before.
>
> —*Ty Churchwell, backcountry coordinator, Trout Unlimited, cited in Quiñones 2015*

THE GOLD KING MINE (GKM) SPILL

On August 5, 2015, the issue of AMD was thrust into the public and political spotlight with the unintended release of 3+ million gallons of subterranean mine water, carrying 880,000 pounds of heavy metals from the entrance of the abandoned Gold King Mine (GKM) into Cement Creek, a tributary to the Animas River in southwest Colorado. Just upstream from its confluence with the mainstem Animas, Cement Creek flows through Silverton, Colorado, the administrative seat of San Juan County. The Silverton area thus became the primary source associated with the spill, where an estimated 120-plus historic mine sites have contributed to AMD for decades (CDRMS 2015). Even prior to the arrival of hardrock mining in the area (circa 1870s), naturally occurring acid rock drainage (ARD) from the underlying geology had degraded water quality for millennia.[4]

Soon after the spill, the entire Animas turned an unusually bright, yellowish-orange color below its confluence with Cement Creek, prompting local officials to restrict public access and suspend multiple municipal intakes and agricultural diversions in Colorado and New Mexico.[5] It took roughly 36 hours for the toxic plume to reach the regional hub of Durango, Colorado, where the Animas has long since been designated a "Gold Medal" fishery by the Colorado Wildlife Commission; it is one of thirteen similarly listed fishing areas across the state's 9,000+ river miles ("Gold Medal Streams" 2018). After crossing into New Mexico, the Animas delivered its discolored plume to the San Juan River, which eventually joins the mainstem Colorado beneath the stagnant waters of Lake Powell (figure 0.1).[6] Throughout the river basin, local communities, Native American reservations, irrigated agriculture, and recreational and wildlife areas were inundated. States of emergency were declared in Colorado, New Mexico, and Utah, as well as by the Navajo Nation Commission on Emergency Management.

Nine days after the initial spill, the toxic plume reached Lake Powell in southeast Utah. All the while, local, national, and international media outlets capitalized on the highly visible, sensational event. The seemingly pristine river in the scenic and diverse corner of the Southwest had turned into a dayglow-orange conduit for acidic, heavy metal–laden water on an inexorable path to the nation's second largest reservoir.

Calls for a strong political response and policy reforms quickly materialized, prompting many to reconsider Superfund listing(s) and anticipate additional changes to existing policies—notably the 1972 Clean Water Act (CWA) and the 1872 General Mining Law. By drawing insights from multiple disciplinary perspectives, this volume adds rich understanding of and context to the dramatic events following the 2015 GKM spill and the ongoing saga of AMD and abandoned mine reclamation across the American West.

As luck would have it, the federal agency in charge of implementing and enforcing many of the nation's most prolific environmental laws accidentally triggered the GKM blowout. The EPA had contracted with a third party

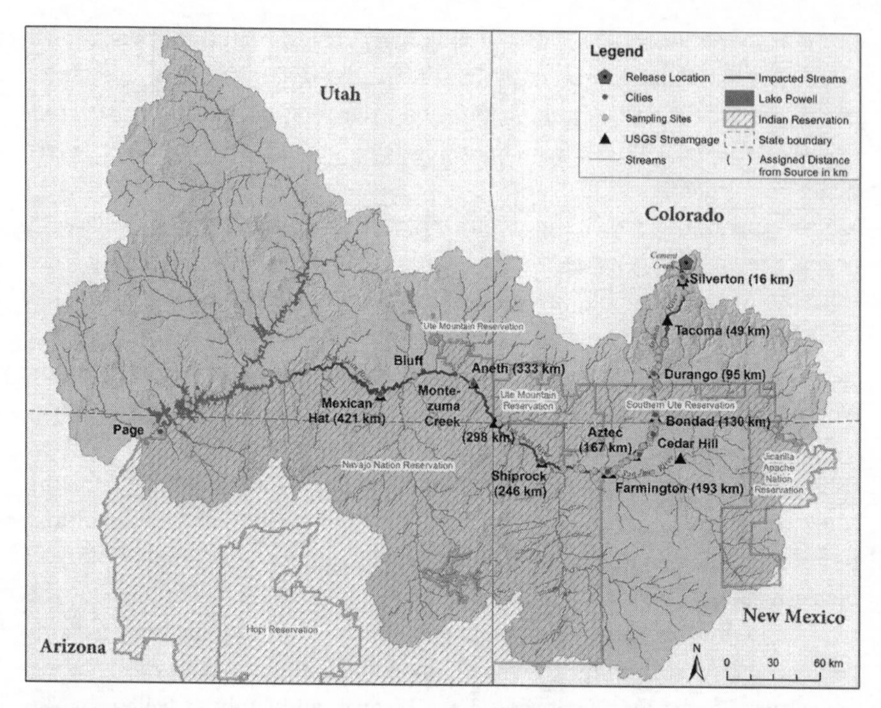

FIGURE 0.1. Path of the GKM Plume along the Animas and San Juan Rivers. *Courtesy*, "One Year After the Gold King Mine Incident: A Retrospective of EPA's Efforts to Restore and Protect Impacted Communities," US Environmental Protection Agency, last updated August 1, 2016.

(Pennsylvania-based Weston Solutions, Inc.) to perform exploratory excavation work to investigate conditions at GKM and assess its ongoing AMD releases. The EPA was quick to assume full responsibility for the spill, and Administrator Gina McCarthy made multiple visits to affected areas, extending apologies for her agency's failed actions.

In a region where local distrust of the EPA is common and opposition to federally led cleanup of abandoned mines has been long-standing, McCarthy's regret for her agency's actions was met with mixed reactions. Some longtime area residents even suggested that the EPA's actions were deliberate and intended to *force* federal cleanup on the Upper Animas River watershed, effectively ending any future mining operations in the region. According to one longtime Silverton resident, "I'm afraid of the EPA. They're too powerful. There's suspicion on my part that now the EPA is sitting judge and jury to decide the outcome of a fate that is a result of their negligence" (Olivarius-Mcallister 2018).

IN THE SPILL'S AFTERMATH: RAPID AND FUNDAMENTAL POLICY CHANGE

A classic view of policymaking in American government described the process as "muddling through," to characterize the behavior of elected officials and public administrators as slow, cautious, and deliberate (Lindbloom 1959). The result is an iterative process, whereby policymaking is (and should be) wholly incremental. Rapid policy development or reversal is considered the exception, not the norm. Hence, analysis of the policymaking process involved "the science of muddling through" (Lindbloom 1959).

Another adage, commonly used in the social sciences, is that *all politics are local*—in the sense that a community-level understanding of issues, events, and problems is essential for understanding policy developments at the national scale. The GKM spill is a case in point; it occurred in the relatively small and isolated Upper Animas watershed yet spawned a national and international media sensation and ensuing debates about the dangers of abandoned mines and AMD.

The strong local opposition to federally led cleanup efforts that had persisted for decades in the Silverton area quickly changed following the spill's visibly disturbing aftermath. After 25-plus years of opposition, it took less than 4 months (or 110 days) for local leaders to vote unanimously to direct city staff members to pursue a Superfund listing with the EPA and the Colorado Department of Public Health and the Environment (CDPHE). Around four months (or 136 days) later, the EPA officially proposed Superfund listing in the *Federal Register* for what would soon become the Bonita Peak Mining District (BPMD). This significant policy development was formalized on September 6, 2016, when the EPA announced the official designation of the BPMD on the NPL—only 137 days after the initial proposal. The time that elapsed between BPMD's proposed and formal listings represents the shortest involved for all of Colorado's nineteen, currently listed, nonfederal NPL facilities; the average time interval is more than 14 months.[7] Anecdotal evidence suggests that this time frame—between proposed and official listing—is a function of the extent and tone of public comments opposing NPL listing (i.e., fewer opposing comments correlates with a shorter time frame).[8] Roughly 60 percent of the public comments submitted by mostly local and regional interests supported the EPA's proposed listing of the BPMD in April 2016.[9]

More broadly, the complete time frame between the initial 2015 spill and formal site listing—a mere 383 days, is remarkable given the decades-long opposition by local leaders and area residents. Such a swift and complete reversal of policy preference is uncommon in American politics, where deliberation and incrementalism (i.e., muddling through) are the norm.

CONTENT AND OUTLINE OF THE WORK

As an editor and author of multiple chapters in this volume, my academic background is in political science and policy analysis. Throughout my undergraduate and postgraduate training, as well as my professional career, I have focused largely on environmental issues, particularly water policy and natural resource management. This has required me to incorporate and expand into my teaching and research aspects from an array of other disciplines—ranging from ecology and geology to history and law. All contributing authors to this work share similar multidisciplinary interests and skillsets, and the majority currently serve as affiliate faculty in the multidisciplinary Environment and Sustainability Department at Fort Lewis College (FLC) in Durango, Colorado. The result of our collaborative efforts is this volume, a uniquely inter- and trans-disciplinary examination into the 2015 GKM spill. Each chapter reflects the professional and personal experiences of its author(s); this allows for a singular event to be surveyed and interpreted from multiple, diverse perspectives.

Our intended audience is similarly broad and diverse; chapters were written with both academic and nonacademic readerships in mind. While all chapters were robustly researched and composed via various academic traditions, deliberate efforts were taken to minimize technical and discipline-specific jargon. The volume is thus relevant for readers broadly interested in hardrock mining in the American West and the legacies of AMD. Chapters should also appeal to readers with more specific interests in any number of other substantive areas, including the history of mining and mining communities in the San Juan Mountains; the region's unique geography, geology, and ecology; environmental law and policy; demographics, socioeconomics, and politics in the Upper and Lower Animas River watersheds; post-spill psychological, economic, and legal impacts; implications for Native American communities, including environmental justice concerns; intergovernmental response to disaster; environmental reclamation strategies; and the potential of future policy developments following the 2015 spill.

The heart of this volume consists of ten chapters written by FLC faculty from eight academic programs, as well as a scientist from a not-for-profit information center based in southwest Colorado. In two chapters (6 and 8), the lead authors recruited as coauthors select FLC students, community activists and educators, and faculty from the University of Arizona's College of Agriculture and Life Sciences. Three of these coauthors are members of the Navajo (Diné) Tribe and thus added valuable perspectives and knowledge to the narrative.

Chapter 1 presents a broad overview of the region's geography, human and hardrock mining histories, and past and present demographic profiles. It was written by a political scientist (Dr. Brad Clark). Chapter 2 discusses the aquatic ecology of the Animas River in both pre- and post-spill contexts. It was written by the water programs director for the Mountain Studies Institute (Scott Roberts).[10] Chapter 3 discusses details of the actual spill and those in the immediate aftermath. Of particular interest is the role of the GKM spill as a powerful focusing event (i.e., an unexpected and dramatic occurrence) and how this prompted profound and unusually fast-paced policy change regarding abandoned mine reclamation in the Upper Animas watershed—official Superfund listing was set in motion a mere 110 days after the GKM event when Silverton officials *unanimously* approved pursuing the federal designation in November 2015. It was written by a political scientist (Dr. Brad Clark). Chapter 4 addresses a host of hydrogeologic and ecological dimensions of the Animas River watershed from the perspective of the natural and physical sciences. It was coauthored by a biologist (Dr. Cynthia Dott) and two geologists (Drs. Gary Gianniny and David Gonzales).

Chapters 5–8 were written from perspectives within the social and behavioral sciences. Chapter 5 places the GKM spill in the context of other major, historic events in the watershed and discusses the central role the Animas River has played in the development of Durango's landscape and sense of place. It was written by a geographer (Dr. Pete McCormick). Chapter 6 examines a range of economic impacts associated with the 2015 spill. It was authored by a professor of management in the School of Business Administration at FLC (Dr. Lorraine Taylor) and her student (Keith Winchester). Chapter 7 discusses the psychological reactions to the GKM spill. It was written by a team of psychologists (Drs. Brian Burke, Alane Brown, Betty Dorr, and Megan Wrona). Chapter 8 examines a host of social and cultural impacts from the spill on communities in the Animas and San Juan River basins. It was written by a group led by a sociologist (Dr. Becky Clausen), along with a hydrologist and environmental engineer (Dr. Karletta Chief), community activists and educators (Teresa Montoya, Janene Yazzie, Jack Turner, Lisa Marie Jacobs, and Ashley Merchant), and a recent FLC graduate (Steven Chischilly).

The next two chapters return to the realm of environmental policy and regulation of hardrock mining. They assess the ongoing development of so-called Good Samaritan legislation, intended to relieve nongovernmental citizen groups from liabilities when initiating AMD remediation projects. Chapter 9 expands on this through a critical examination into the problems associated with court litigation as a means to ensure implementation and enforcement of federal environmental laws. It was written by a political scientist (Dr. Michael Dichio). Chapter 10 examines the primary actors behind the two competing perspectives regarding AMD remediation in San Juan County—both prior to and immediately following the 2015 GKM spill. It was written by a political scientist (Dr. Brad Clark). Finally, in the afterword, historian Dr. Andrew Gulliford employs the saying *we all live downstream* to highlight the many lessons to be learned from the 2015 GKM spill.

REMAINDER OF THE INTRODUCTION

This chapter concludes with a brief history of gold and silver mining in Colorado and, specifically, historical activities in the Cement Creek drainage. The AMD problem is then defined and its geologic and *anthropogenic* (i.e., human-induced) causes are discussed. The chapter ends with a brief chronology of events before and immediately after the spill. An update on the most recent developments (circa August 2018) in the unending GKM story is included.

TABLE 0.1. Time line of significant GKM-related events

Year	*Event*	*Description*
1860	Baker Party arrives in Upper Animas watershed.	Discovery of gold and silver deposits in Baker's Park area near present-day Silverton.
1860–1861	Animas City established.	Becomes first trading hub in the lower Animas River Valley.
1873	Brunot Agreement.	United States assumes control of 4 million acres from Utes.
1874	First mining rush; Sunnyside Mine patented.	An estimated 2,000 prospectors establish 1,000 mining claims in Baker's Park area.
1876	Colorado statehood; Town of Silverton incorporated.	Silverton poised to become mining hub of San Juan County.
1877	Animas Canyon Toll Road completed, linking Silverton to lower Animas Valley.	Increased delivery of supplies and materials to miners in Silverton area; transport of ores to Animas City (eventually Durango).
1880	Durango incorporated.	Durango will subsume Animas City by 1940s.
1881	Denver & Rio Grande Railroad reaches present-day Durango.	Animas City (later Durango) established as regional hub.
1882	Railroad reaches Silverton.	Rail linkage provides foundation for growth.
1887	Gold King Mine (GKM) established.	Olaf Nelson stakes claim; never becomes rich; dies of pneumonia 4 years later.
1890–1920	Primary production era at GKM.	665,000 tons of ore (silver, gold, lead, copper) produced.
1894	GKM sold for $15,000.	First of many changes in ownership.
Early 1940s	Uranium processing in Durango.	Uranium milled for Manhattan Project, atomic weapons.
1963	Uranium processing ends.	Nearby lands and Animas severely contaminated.
1975	Mine tailings spill in Silverton.	Roughly 50,000 metric tons spilled into Animas River.
1985	Lake Emma disaster.	Lake above Sunnyside Mine collapses into mine tunnels; 500 million gallons flood mine, AMD blowout.
1985–1991	Superfund cleanup (Durango).	Site remediation and relocation of radioactive wastes.
1994	Animas River Stakeholders Group (ARSG) forms.	ARSG starts local AMD remediation projects, becomes outlet for opposition to Superfund.
1996–2002	American Tunnel bulkheads installed.	AMD from Sunnyside decreases; AMD from GKM and others in Cement Creek drainage increases significantly.
2009	Annual AMD from GKM at 200,000 lbs. of heavy metals.	GKM labeled one of the worst AMD sources in Cement Creek by Colorado Division of Reclamation, Mining, and Safety.

continued on next page

TABLE 0.1—*continued*

Year	*Event*	*Description*
2014	Sunnyside Gold Corporation permanent treatment plant.	$10 million offer has stipulation that if accepted, potential Superfund listing in Cement Creek permanently stopped.
2015	GKM blowout, 10:30 a.m. on August 5.	3 million gallons of AMD released into Cement Creek.
2015	36 hours later, plume hits Durango.	Animas River turns orange in color; river and all intakes closed; intense media coverage.
2015	Silverton, San Juan County pursue Superfund listing.	Unanimous approval by vote on November 23.
2016	Colorado governor requests BPMD listing.	Governor Hickenlooper requests adding BPMD to NPL on February 29.
2016	EPA proposes Superfund listing.	EPA proposes BPMD listing on April 7
2016	Final Superfund listing by EPA.	BPMD announced on September 9 in the Federal Register.
2016	First of multiple lawsuits on May 23.	New Mexico sues EPA, mine owners; New Mexico sues Colorado on June 30; Navajo Nation sues EPA, mine owners on August 17.
2018	EPA outlines cleanup plan (June).	26 sites to be restored via "Interim Remedial Action Plan."

Hardrock Mining in Colorado

Following the California Gold Rush, gold fever came to Colorado in June 1858, when prospectors began sluicing sand and gravel at the confluence of Cherry Creek and the South Platte River near the present-day location of downtown Denver. Soon thereafter, 100,000 prospectors followed, and by the early 1860s, many had found their way south and west to the Upper Animas River watershed in the state's rugged San Juan Mountains. The area, which would become San Juan County following Colorado's 1876 statehood, quickly became one of the most important hardrock mining regions in the Rocky Mountain West (US DOI 2015).[11]

Cement Creek and the Gold King Mine

The Upper Animas watershed draws from three main sources—the Animas headwaters, Cement Creek, and Mineral Creek—all of which have historically been impacted by AMD. In the Cement Creek drainage, a cluster of abandoned sites at and above the historic mining ghost town of Gladstone has been the primary source of metals loading. Along with the GKM, these mine sites include the American Tunnel, the Red and Bonita Mine, and the Mogul Mines.

At an elevation of approximately 11,400 ft., the GKM was established in 1887, when Olaf Nelson first staked a claim high in Cement Creek's north fork on the slopes of Bonita Peak. Following multiple changes in ownership, the mine continued to operate until late 1922 and has remained largely out of operation since 1923. During its time, GKM produced 711,144 tons of gold and silver ore from seven separate levels, spanning 760 vertical feet ("Gold King Mine" 2015). At the time of the 2015 blowout, GKM was owned by Todd Hennis of the Golden, Colorado–based San Juan Corporation. Hennis also owns the neighboring Mogul Mine in the Cement Creek drainage.

GKM is located within an extensive volcanic field in the Upper Animas watershed, in what is commonly referred to as the collapsed Silverton caldera. The area represents the southern terminus of the Colorado Mineral Belt, which runs diagonally across the state from Durango to Boulder. Because of its volcanic origins and underlying geology, the Upper Animas watershed is naturally a highly mineralized region, which has experienced several heat-induced mineralizing events over the previous 25 million to 35 million years. These processes deposited valuable metals at GKM and altered surrounding rocks ("Gold King Mine" 2015). As a result, Cement Creek and surrounding waterways receive acid drainage that is naturally formed, albeit less concentrated and more dispersed. Indeed, select tributaries to the Animas—including Cement Creek—were practically devoid of life (i.e., biologically dead) prior to the onset of mining activities.[12]

During the heyday of hardrock mine production (circa 1890–1920), an estimated 4.3 million tons of tailings were discharged directly into Silverton-area streams via the many large gold and silver mills, such as the one formerly located at the Gladstone townsite (Church et al. 2007). By the early 1900s, downstream conditions in Durango had deteriorated to the point that the city was forced to switch its municipal supply from the Animas to the Florida River watershed, an area with a far less extensive history of hardrock mining. In Silverton, drinking water has long since been sourced from an Animas tributary well above Cement Creek.

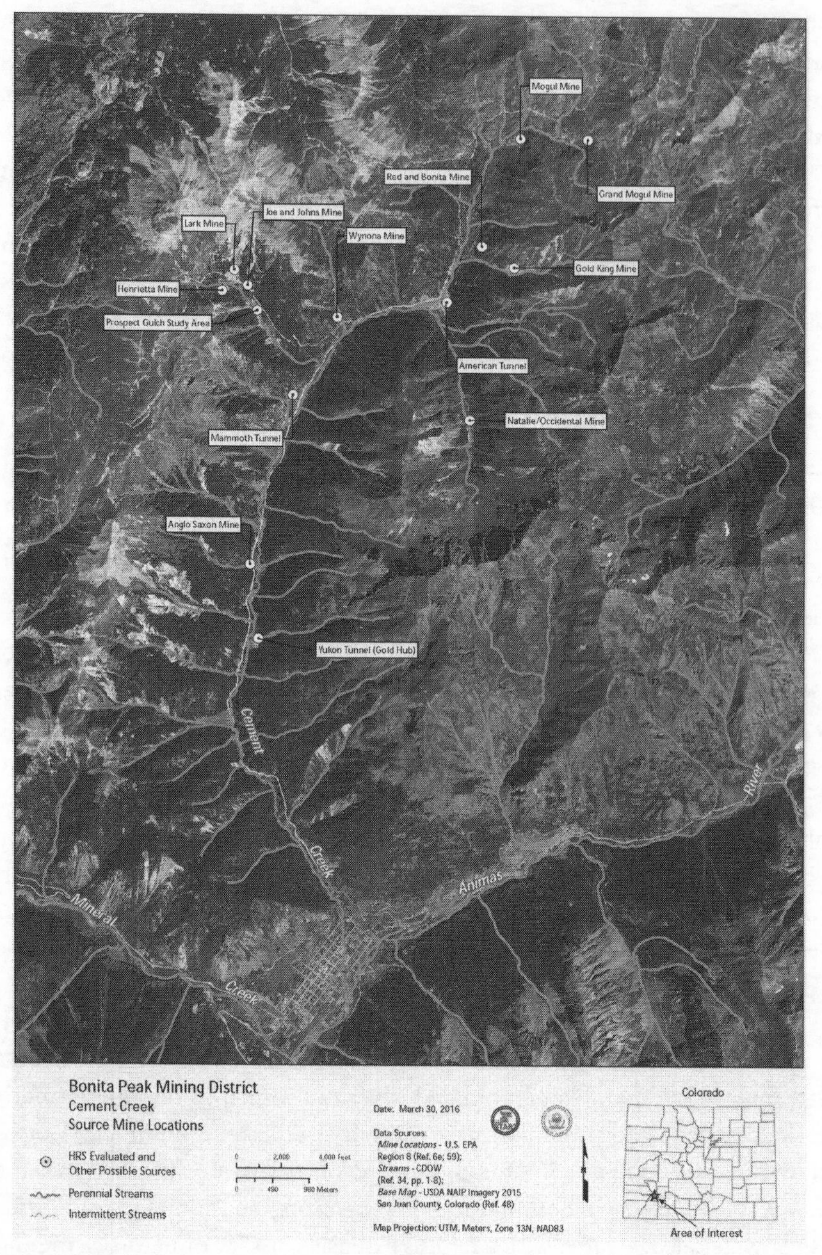

FIGURE 0.2. Mines in the Cement Creek drainage. *Courtesy*, "Gold King Mine Release—Analysis of Fate and Transport in the Animas and San Juan Rivers," US Environmental Protection Agency, last updated June 21, 2016.

It wasn't until the mid-1970s that laws were passed to protect the environment from the impacts of hardrock mining. Without regulation or the required posting of bonds to ensure reclamation of mined sites, prospectors could freely disturb the landscape and impact waterways; when mining activities ended, they could simply and *legally* walk away.

With thousands of mines having been sunk into mountainsides, precipitation and discharge from natural springs perpetually accumulate in the many miles of subterranean tunnels and shafts.[13] The waters react with naturally occurring iron sulfide minerals (e.g., pyrite) and oxygen, which produces sulfuric acid. These acidic waters dissolve the area's naturally occurring heavy metals, including zinc, arsenic, lead, cadmium, copper, aluminum, thallium, and selenium. After periods of accumulation, metal-laden waters inevitably discharge from the hundreds of mine adits (or openings).[14]

In addition, pyrite is the most common material in the area's piles of mine tailings. When exposed to oxygen and precipitation at the surface, tailings further contribute to the production of acidity and sulfides. Quite simply, AMD is produced virtually whenever and wherever pyrite is exposed to oxygen and water. Along with mine adits, this makes tailings and waste piles primary culprits of anthropogenic AMD. In total, since mining began in the watershed, an estimated 8.6 million tons of tailings ended up in the Upper Animas River environment ("Technical Evaluation" 2015).

In Cement Creek, AMD has caused its pH value to fall to around 3.5, which is similar to that of store-bought vinegar. In such acidic water, heavy metals are soluble (or easily dissolved). After being diluted with less acidic waters (e.g., the mainstem Animas), pH levels rise; as this happens, metals begin to (re)solidify in the water column and eventually settle as contaminated sediments. The orange-ish color of the impacted waters is the result of these heavy metals (e.g., copper and zinc) becoming attached to iron particles. Together, these processes have had tremendous impacts on aquatic ecosystems in nearby waters. For example, no fish have been found to survive in the Animas for approximately 2 miles after it is joined by Cement Creek, and precipitous declines in fish populations have been reported as far as 20 miles downstream from this confluence (US EPA 2016). For communities serviced with drinking water from contaminated rivers and streams, the bioaccumulation of metals is a public health concern.

Before the 2015 GKM incident, there were two significant AMD releases in the Upper Animas watershed. In June 1974, a tailings dam breached at Sunnyside Mine's mill in Silverton and an estimated 116,000 tons of acidic tailings were released into the Animas. As a result, the cities of Durango, Colorado, and Farmington, New Mexico, were forced to close municipal water intakes for a period of seven days and fish kills were reported near Durango, 40 miles away from the spill (Bird 1986). A second release occurred in June 1978, when a portion of Sunnyside Mine located beneath Lake Emma collapsed, releasing 500 million gallons of sediment-laden water into the mine's tunnels. Contaminated waters soon burst via the mine's American Tunnel, launching wrecked mine equipment, timbers, and sulfide rock tailings into the Animas (Bird 1986).[15]

Large-scale mining in the area ended in 1991, when the Sunnyside Mine and its American Tunnel were closed due to a combination of declining ore reserves, falling gold and silver prices, and mounting concerns over water quality in Cement Creek.[16] In fact, a year prior to Sunnyside's closing, the State of Colorado's Water Quality Control Division (CWQCD) had begun a program to establish water quality standards in the Upper Animas watershed. Preliminary analysis indicated that the majority of heavy metals were coming from around eighty abandoned mine sites across the Upper Animas watershed. A number of these were initially targeted for reclamation to minimize drainage and restore impacted surface lands. In general, the projects involved installation of hydraulic bulkheads (i.e., concrete plugs) at mine openings—to stem AMD that had accumulated deep within the mines, as well as construction of settling ponds and basic treatment plants—to remediate residual seepage and increase pH levels through the addition of lime.[17] GKM was not included in these initial cleanup efforts, as other sites received priority.

Releases of contaminated water decreased notably at many of the sites where bulkheads were installed and treatment facilities constructed. However, impounded waters at many of these sites simply migrated to neighboring subterranean voids or fissures and steadily accumulated in neighboring mines. By most accounts, GKM was chief among these in the Cement Creek drainage; shortly after the first Sunnyside Mine bulkhead was completed in 1996, seepage from GKM increased as discharge from Sunnyside *dropped* from 1,700 to 100 gallons per minute (gpm).

Two additional bulkheads were completed at the Sunnyside Mine and American Tunnel between 2001 and 2002; and in 2003, bulkheads were added to the Mogul Mine. Again, significant increases in discharge were detected from GKM and other neighboring mines. These waters were initially stored and remediated at a treatment plant near Gladstone, which by 2004 was closed due to a host of technical, financial, and legal troubles. All the while, additional waters pooled via drainage from tunnels behind Sunnyside's bulkheads, and discharges from GKM and others steadily rose.[18] By 2006, peak discharges from GKM had risen to over 300 gpm. Three years later, GKM was found to be releasing nearly 200,000 pounds of metals into the watershed each year; this led to its ranking as "arguably one of the worst high quantity, poor water quality draining mines in the State of Colorado" by the CDRMS (Thompson 2015). Also in 2009, the CDRMS closed all of the adits at GKM by backfilling their portals with various fill materials; they were not bulkheaded. Subsequent to the backfilling, AMD averaged 200 gpm from GKM's lowermost Level-7 adit.

By 2010, significant declines in brook trout populations in the Animas River below Silverton had been reported by the Colorado Department of Wildlife. Between 2010 and 2014, AMD discharge from GKM averaged 153 gpm. In part, this prompted the EPA to initiate additional data gathering on water quality as an initial step in the process of determining whether Cement Creek was eligible for federally led reclamation under Superfund. The EPA also began petitioning Silverton's elected officials and area residents to consider supporting Superfund designation.

As the Superfund option gained traction over the next few years, Hennis and the Sunnyside Gold Corporation (SGC) made repeated offers to fund construction of a permanent water treatment facility for the Cement Creek drainage as the Sunnyside Mine and American Tunnel continued to discharge AMD at the rate of 100 gpm, despite the multiple sealed bulkheads. In exchange for its $10 million overture, SGC asked for total exemption from any potential liabilities for future restoration efforts in Cement Creek. Such an offer was hardly surprising, given that SGC is among the largest *potentially responsible parties* (PRP) under Superfund financially liable for comprehensive AMD remediation in the Upper Animas watershed. In addition, the offer was backed by SGC's parent company—the Kinross Gold Corporation, an international, publicly traded mining conglomerate based in Canada.[19]

Nothing ever materialized, as the $10 million offer was left on the table until the GKM blowout occurred.

In 2014, the EPA returned to the GKM site at CDRMS's request that it reopen and stabilize the Level-7 adit that had previously collapsed. The drainage system and backfilling at the entrance had reportedly not been maintained or routinely monitored since 2009 ("Technical Evaluation" 2015). The EPA quickly determined that additional time and resources would be necessary to complete GKM's reopening, and work stopped after a few hours of excavation. The remaining work was postponed until the following year, when bulkhead construction was completed by early summer 2015 at the nearby Red and Bonita Mine. Its valve, however, was left open out of concern that closing it could increase water levels at GKM.

By late July 2015, EPA contractors had returned and resumed work by reconstructing the access road to GKM. On the morning of August 4, an EPA on-scene coordinator and an official from the CDRMS arrived at GKM. Per their instructions, EPA's Emergency and Rapid Response Services contractor began to excavate an area at the mine's collapsed Level-7 adit. The drainage from GKM was measured at 69 gpm. After less than 4 hours of work, a set of collapsed timbers was uncovered at GKM's presumed opening. Excavation stopped, and workers left the site altogether to allow time for overnight consideration of future activities. And then it happened . . .[20]

On the morning of August 5, 2015, it was determined that excavation should continue at the Level-7 adit. At 10:51 a.m.—roughly 80 minutes after work began—a small leak of water appeared 15 feet to 20 feet above the adit's floor. The EPA's contractor had allegedly miscalculated the depth and pressure of water accumulated behind the bulkhead (US House Committee on Natural Resources 2016). Within minutes, a portion of bedrock fell away from the mine's opening and a greater volume shot upward 1.5–2-feet; the breached opening later measured 10 feet in width and 15 feet high. Initially, the discharging water was clear but soon changed to a reddish-orange color. The excavator's operator reported that he had hit a "spring" and quickly removed the machine.

The access road was soon destroyed by the burst of water, and the EPA's vehicle was rendered undriveable. It took approximately an hour for the peak flow to subside, after which discharge decreased to a fairly constant 500–700 gpm. The pH was measured (by a handheld paper test) at 4.5—a strongly acidic level similar to that of black coffee.

The onsite workers had no cell phone or satellite connections; nearly an hour elapsed before they were able to establish radio contact with offsite personnel, who subsequently issued initial notifications of the spill to the EPA's on-scene coordinator and the CDRMS. It took almost another hour for the state's main regulatory agency with direct jurisdiction over the accident—the CDPHE—to complete notifications to the City of Durango, the San Juan Basin Health Department, and other operators of water intakes.

Initial water quality samples were not taken until roughly 6:00 p.m.; after two days, results indicated elevated levels of copper, lead, manganese, and zinc. By August 8, GKM's discharge flow rate had steadied to an approximate average of 587 gpm and its pH had fallen to 3—similar to that of orange or grapefruit juice.[21] A week after the initial spill, CDPHE reported that metals loading in the Animas had returned to pre-spill levels, while levels of cadmium, copper, and zinc remained above historic standards in Cement Creek.

ISSUES AND DEVELOPMENTS FOLLOWING THE SPILL

By May 2019, almost four years after the spill, a number of notable issues and important developments had impacted the course of the ongoing GKM story. For context and a brief update, the following warrant retention.

Water Treatment

After official Superfund listing of the BPMD, a $1.5 million temporary water treatment plant was constructed on Cement Creek near the historic mining hub of Gladstone, roughly 8 miles north of Silverton. It became operational in October 2015. At the plant, lime is added to the AMD to raise overall pH levels in Cement Creek. This causes dissolved metals to solidify in the water column and settle in retaining ponds. The practice is generally effective, yet a tremendous amount of sludge is generated in the process—an estimated 4,600 cubic yards per year from the average 450 gpm of discharge from GKM. Further, it will likely be necessary for the treatment plant to be operated *in perpetuity*.

On the evening of March 14, 2019, a period of unusually heavy snowfall caused a power outage at the treatment plant as well as an avalanche that temporarily blocked access to the plant. After a period of less than 48 hours, the EPA brought the plant back online, and it resumed normal operations on the afternoon of March 16. During shutdown, an estimated 264 gallons per minute of untreated AMD drained into Cement Creek and the Animas River. As a precautionary measure, the water intake facilities for the Cities of Aztec and Farmington, New Mexico, were temporarily closed. The EPA conducted water quality samples from four locations along the Animas River, from its confluence with Cement downriver to sites above and in Durango. A considerable elevation of heavy metals, especially copper, was detected at the confluence and lower concentrations in the Animas roughly 1 mile downriver of Silverton. The two sampling locations in Durango yielded heavy metal concentrations within the range of those measured when the water treatment plant was operational.

Sludge Storage, Transport, and Safety

When the treatment plant went online in October 2015, the sludge was stored onsite along Cement Creek at an area known as Gladstone. According to the EPA, the amount of available storage space was to be entirely filled by August 2018. For months previous, the EPA had unsuccessfully searched for additional storage space in the Upper Animas watershed. The massive tailings ponds north of Silverton, operated by SGC (GKM's current owner), were identified as suitable locations, yet the company was listed in 2018 as a PRP for EPA's cleanup and the two sides were unable to agree on acceptable uses for the ponds. As the one remaining viable option, it was decided that all future sludge would be transported by truck more than 70 miles and over two mountain passes to a landfill south of Durango, near the New Mexico border. Elected officials and the majority of residents in both San Juan and La Plata Counties opposed the plan on financial, public safety, and environmental grounds—particularly the carbon footprint that would result from the estimated 700 annual trips spanning an unknown number of years.

With time running out on the search for an alternative plan, the EPA reached an agreement for onsite storage at the historic Kittimac tailings pile, a 10–15 mile one-way drive from the Gladstone treatment site. Thus the short-term problem of sludge storage has been addressed, but what remains is the larger, more complex issue—what to do with the massive amounts of sludge that will result from long-term AMD treatment under Superfund.

Meanwhile, in the near term, safety concerns over sludge transport across *any* distance remain valid. A few days into the trucking of sludge to the Kittimac site, a vehicle driven by an EPA contractor crashed into Cement Creek, releasing roughly 9 cubic yards of sludge material into the waterway. The driver was not seriously injured, yet such an inauspicious start hardly inspires much confidence.

Mine Spill versus Forest Fire: Comparing the Impacts Following Disaster

Roughly 3 years after the GKM spill, the Animas River Basin played host to another catastrophic event—massive wildfires. On June 1, 2018, the 416 Fire started 10 miles north of Durango; a week later the Burro Fire ignited a short distance to the northwest. Together, the officially named 416 and Burro Complex Fire would burn more than 54,000 acres, making it the sixth largest and most destructive fire in Colorado history.

The fire prompted the first-ever official closing of the nearly 1.9 million acre San Juan National Forest, which included an extended suspension of operations of both the Purgatory Resort and the Durango & Silverton Narrow Gauge Railway, which normally carries up to 193,000 tourists per day during the summer months and injects $190–$200 million annually into the two tourist towns (Best 2018). With the tourist train suspended, the economy in San Juan County was particularly decimated.

In contrast, the economies of Durango and La Plata and San Juan Counties showed minor and only temporary negative impacts following the GKM spill. Despite a reliance of up to 20 percent on tourism, many economic sectors across the area—including lodging, retail, and food and beverage—actually posted higher than average sales for August 2015. In Durango proper, sales tax revenues for August 2015 were up 2.5 percent.

The fire's ecological footprint was similarly enormous, with longtime impacts far in excess of those following the 2015 GKM spill. In particular, flash floods and debris flows from the 416 burn scar led to massive fish kills in the Animas River and its tributary, Hermosa Creek. These were attributed primarily to suffocation that resulted from high ash concentrations and low dissolved oxygen levels in the waterways. Fish surveys by the CDPW found fish populations (both brown and rainbow trout) in multiple river and creek segments that were more than ten times below historical levels.

Concentrations of heavy metals were also much higher following the 2018 fires than those detected during the flow of GKM's toxic plume through Durango. While no evidence of any die-offs of aquatic life was linked to the GKM's AMD, heavy metal concentrations caused a near-total fish kill in the Animas after the 2018 fire season. Specifically, concentrations of aluminum were measured at fifty times higher than those associated with GKM's plume; iron levels were six times higher; manganese was twenty times higher; levels of mercury were three times higher than those registered at the peak of the GKM event (Romero 2018). Granted that the fires of 2018 were slow-moving disasters lasting more than two months and the GKM plume discolored the Animas for roughly a week, the degree, magnitude, and duration of media coverage regarding the former far exceeded that of the latter. Yet the political attention and policy change directly following the GKM spill were much more intense and substantive than those resulting from the 416 and Burro Fire Complex.

Initial Remediation Plans

Also in June 2018, the EPA released its proposed Interim Remedial Actions Plan (IRAP)—informally referred to as the "quick-action cleanup plan," for inaugural restoration work at twenty-six abandoned mining sites in the BPMD outside of Silverton (Romero 2018). In particular, the EPA created five types or sources of potential AMD-related pollution (officially termed "contaminant migration issues" [CMI]) at the twenty-six sites. The first involves mine portal discharges, of which there are twenty. The second targets eleven sites where stormwater and mining-related materials commingle. The third focuses on contaminated sediments held in mine portal settling ponds. The fourth centers on two sites where mine wastes are entirely within or located on both banks of a waterway. The final type of contaminant migration issue targets two mining-impacted recreation areas (e.g., dispersed campsites) where tailings piles or contaminated soils levels of arsenic and lead are in excess of human-health thresholds (Romero 2018).

According to the EPA, the CMI-related work will occur simultaneously with the formulation of the more comprehensive long-term plan known as the "Site-Wide Remedial Investigation and Feasibility Study," which will address restoration at all forty-eight mine sites within the BPMD, including

GKM and others in the Cement Creek drainage. Specific to the GKM site, the EPA plans to transition the temporary water treatment plant at Gladstone to some form of a permanent facility by January 2022.

The majority of public comments submitted in response to this proposed "quick-action" plan were critical; many commenters cited their disapproval of the EPA's focus on remediation activities at a large number of relatively minor sites of AMD-related pollution as opposed to focusing on a smaller number of major sites of AMD contamination—specifically, those high in the Cement Creek drainage such as the GKM, the Red and Bonita Mine, the American Tunnel, and the Mogul Mine complex.[22]

In addition, many commenters expressed ongoing skepticism toward and distrust of the EPA's takeover of AMD remediation in the Upper Animas watershed. For example, Peter Butler, then-co-coordinator of the now-defunct ARSG, reacted to the proposed interim plan by stating that the plan "seems to have been developed for the political purpose of showing that something is being done as opposed to developing an overall cost-effective strategy for improving water quality" (quoted in Romero 2018). Reflecting a similar sentiment, Bill Simon, former coordinator emeritus of the ARSG, stated that "if one was serious about 'draining the swamp' I would suggest saving most of the over $8M and 10 years of fiddling around and instead attack the real cause of the problem. That alone will go further than a bunch of high visibility 'feel good' projects primarily designed for PR purposes" (quoted in Romero 2018).

In response, the interim project manager for the BPMD stated that the extreme complexity of the mine network in Cement Creek justified the proposed activities and that years of investigation are necessary for development of a long-term solution for large sources of AMD such as GKM. Specifically, it was stated that the proposed plan deals with "immediate steps that are relatively straightforward and simple" across the entire headwaters of the Animas River (Romero 2018).

In May 2019, the EPA released its final Interim Record of Decision (IROD); it included minor changes to the 2018 IRAP, based primarily on consideration of public comments and ongoing feasibility studies. Specifically, the IROD reduces the number of targets to twenty-three (including the two dispersed campsites) and retains the five CMIs—which were renamed Interim Remedial Actions (IRAs). Together, implementation of the five IRAs at the

TABLE 0.2. Interim Record of Decision (IROD)—Costs and locations (US EPA 2019)

IRA	*Cost (millions)*	*Location(s)*	*Target(s)*
Mine portal Discharge for "mine influenced water" (MIW)	≈ $2.5	18 mining-related sources of AMD	13 mines and 5 tunnels
Mining-related source/ stormwater interactions	≈ $2.0	10 mining-related sources of AMD	9 mines and 1 tunnel
Mine portal pond sediments	≈ $3.7	8 mining-related sources of AMD	5 mines and 3 tunnels
In-stream mine wastes	≈ $0.50	1 mining-related source of AMD	1 mine
Mining-impacted recreation areas	≈ $1.7	5 mining-related sources of AMD	2 mines, 1 tunnel, and 2 campgrounds
Total	≈ $10.4		

Note: The total number of locations listed in the IROD is twenty-three. Many locations and targets are listed multiple times for different IRAs.

twenty-three sites will cost nearly $10.4 million in total, which includes initial construction activities as well as operation and maintenance costs over a 15-year period (US EPA 2019). For a cost breakdown of the IRAs, see table 0.2.

NOTES

1. An adit is a horizontal or near-horizontal opening to an underground mine used for entrance, ore removal, drainage, and ventilation.

2. Anglers are especially affected by AMD and have become a potent force in calling for restoration of abandoned sites. Their main lobbying organization, Trout Unlimited, continues to devote significant resources to heighten awareness and promote legislation intended to facilitate cleanups.

3. The pace of policy development increased in 1976, when the Colorado Mined Land Reclamation Division was created to regulate non-coal mining operations. That same year, the Colorado General Assembly passed the Colorado Mined Land Reclamation Act, which created the Mined Land Reclamation Board to serve in administrative and adjudicatory capacities over non-coal mines.

4. AMD and ARD are formed when pyrite (an iron sulfide, or FeS_2) is exposed to and reacts with oxygen and water to form sulfuric acid (H_2SO_4) and dissolved iron (Fe). Some or all of this iron precipitates to form the red, orange, or yellow sediments in the bottom of streams containing AMD and/or ARD. The acid runoff further dissolves heavy metals such as copper, lead, and mercury into surface water and groundwaters.

5. AMD has caused the pH value in Cement Creek to fall to around 3.5, which is similar to that of store-bought vinegar. In such acidic water, heavy metals are soluble (or easily dissolved). After being diluted with less acidic waters (e.g., from the mainstem Animas), pH levels rise; as this happens, metals begin to (re)solidify in the water column and eventually settle as contaminated sediments. The orange-ish color of the impacted waters is the result of these heavy metals (e.g., copper and zinc) becoming attached to iron particles.

6. The distance from the GKM site on Cement Creek's north fork to the New Mexico state line is 83 miles. The Animas continues an additional 30-odd miles to its confluence with the San Juan River. Since the completion of Glen Canyon Dam in 1963, the San Juan joins the Colorado at Lake Powell.

7. The shortest time between the proposal and official NPL listing in Colorado was 130 days, for the Air Force Plant PJKS (Peter J. Kiewit and Sons) federal site. Cleanup and construction activities were completed by February 2014; the site is currently owned and operated by Lockheed Martin Astronautics Operation. The 464 acre plant is located 25 miles southwest of Denver and is surrounded by another 4,700 acres of Lockheed Martin property. Operations at PJKS included testing Titan rockets, as well as design and manufacture of technical systems for space and defense.

8. For information about stages of NPL listing, see https://www.epa.gov/superfund/superfund-cleanup-process.

9. See public comments made in response to EPA's proposed NPL listing at https://semspub.epa.gov/work/08/1570791.pdf.

10. The Mountain Studies Institute (MSI) is an independent 501(c)3 not-for-profit center of knowledge established in 2002 in Silverton, Colorado. MSI's currently stated mission is "to empower communities, managers, and scientists to innovate solutions through mountain research, education, and practice." For more information, see http://www.mountainstudies.org/mission-vision.

11. In August 2015, the Colorado Division of Reclamation, Mining, and Safety classified 120 Silverton-area mine sites as the following: 18 mines had active water treatment, 18 mines were under investigation, 66 mines had no active treatment, and 18 nonpoint sources (i.e., waste piles of ore) had no active treatment. These AMD sources were classified as "actively" or "likely impacting" water quality. See http://mining.state.co.us/Programs/Abandoned/Documents/LegacyMineWork.pdf.

12. Coincidentally, the naturally acidic waters of Cement Creek support one of the few areas on the planet where iron fens are located. Iron fens are a type of unique wetland that has highly acidic water (a 4.5 pH or less), which results in a diverse array of plant life. Most iron fen plant species (e.g., Sphagnum mosses) are

found only in the boreal forests of Canada and Alaska, more than 1,200 miles away. Currently, there are only thirteen iron fens globally; four of these are located in San Juan County. "Appendix D Wild and Scenic Rivers Suitability, last modified September 2013," https://www.fs.usda.gov/Internet/FSE_DOCUMENTS/stelprdb5435197.pdf.

13. GKM is in an alpine environment and experiences significant precipitation, mostly in the form of snow from late fall to spring. Snow accumulations average roughly 15 feet.

14. An adit is a horizontal or near-horizontal entrance to an underground mine, by which the mine can be entered, drained, and ventilated and minerals extracted.

15. The event happened on a Sunday, when operations at the Sunnyside Mine were down. Had it occurred on a workday, a crew of 125 miners would have likely perished from the sudden inrush of water into the mine and the subsequent blowout into the Animas. See Bird (1986, 16).

16. The American Tunnel is often mistaken for an actual independent mine. In actuality, it is the lower level of the Gold King Mine at Gladstone, Colorado. In 1959, the American Tunnel was extended more than a mile to intersect Sunnyside Mine, 600 feet below the original mine workings. As such, the American Tunnel is Sunnyside's lowest transportation and ore-haulage level. See Rosemeyer (2017).

17. An alternative to the practice of bulkhead placement at mine openings in order to block AMD has been referred to as "source control," whereby hydraulic controls (e.g., rock plumbing) are installed above mine adits to prevent precipitation and other surface flows from infiltrating the inner workings of abandoned mines. The goal is to minimize or prevent water from contacting mineralized deposits. Such a source control approach has been implemented to remediate zinc pollution at the Idarado Mine in neighboring Ouray County. See Fiscor (2015).

18. Discharge from the Red and Bonita Mine contains extremely high levels of zinc, while drainage from GKM is much more acidic. See US DOI (2015).

19. For 2017, Kinross posted a net profit of $445.40 million.

20. This chronology of events was derived from two sources: US EPA (2015); EPA Region 8 (2015).

21. The pH scale runs from levels 1–14, with lower values indicating higher acidity. Each whole number up or down in the scale is equivalent to 10 times the concentration of the previous step. Neutral water has a pH of 7, but a pH of 6–8 is generally considered normal. Depending on the species, the recommended pH for fish is between 6.0 and 9.0.

22. The 30-day public comment period lasted from June 14 through July 16, 2018.

REFERENCES

Best, Allen. 2018. "Coal Gives Way to Diesel as Fire Danger Rises." *Mountain Town News* (Arvada, CO). http://mountaintownnews.net/2018/08/04/coal-gives-way-to-diesel-on-narrow-gauge-railroad/.

Bird, Allan G. 1986. *Silverton Gold: The Story of Colorado's Largest Gold Mine*. Silverton, CO: Paul and Meridelle Bird.

CDRMS (Colorado Division of Reclamation, Mining, and Safety). 2015. "Preliminary Maps of Legacy Draining Mine Adits." http://mining.state.co.us/Programs/Abandoned/Documents/LegacyMineWork.pdf.

Church, Stanley E., J. Robert Owen, Paul von Guerard, Philip L. Verplanck, Briant A. Kimball, and Douglas B. Yeager. 2007. "The Effects of Acidic Mine Drainage from Historical Mines in the Animas River Watershed, San Juan County, Colorado—What Is Being Done and What Can Be Done to Improve Water Quality?" In *Understanding and Responding to Hazardous Substances at Mine Sites in the Western United States*, edited by Jerome V. DeGraff, 57–58. *Reviews in Engineering Geology XVII*. Boulder: Geological Society of America.

EPA Region 8. 2015. "Gold King Mine Investigation and Blowout Event." Memorandum 1574032. On-Scene Coordinator to Superfund Technical Assessment and Response Team (START), August 12.

Fiscor, Steve. 2015. "Gold King Spill Daylights EPA's Poor Remediation Practices." https://www.e-mj.com/features/gold-king-spill-daylights-epa-s-poor-remediation-practices/.

"Gold King Mine—Watershed Fact Sheet." 2015. Washington, DC: US Environmental Protection Agency. https://www.epa.gov/sites/production/files/2015-08/documents/goldkingminewatershedfactsheetbackground.pdf.

"Gold Medal Streams in Colorado." 2018. *Colorado Fishing*. http://coloradofishing.net/goldmedal.htm.

Limerick, Patricia N., Joseph N. Ryan, Timothy R. Brown, and T. Allan Comp. 2005. *Cleaning up Abandoned Hardrock Mines in the West: Prospecting for a Better Future*. Boulder: Center for the American West, University of Colorado.

Lindblom, Charles. 1959. "The Science of 'Muddling Through.'" *Public Administration Review* 19 (2): 79–88.

Moyer, Steve. 2016. "Discussion Draft of Good Samaritan Cleanup of Orphan Mines Act of 2016." Testimony, US Senate Environment and Public Works Committee, Washington, DC, March 2.

Olivarius-Mcallister, Chase. 2018. "Is Silverton Ready for a Cleanup?" *The Journal* (Cortez, CO). https://www.the-journal.com/articles/23931.

Parrington, Vernon L. 1930. *The Beginnings of Critical Realism in America 1860–1920: Main Currents in American Thought*, vol. 3. New York: Harcourt Brace.

Quiñones, Manuel. 2015. "EPA's Spill Pales in Comparison to Everyday Mine Leaks." *EE News*. https://www.eenews.net/stories/1060024348.

Romero, Jonathan. 2018. "EPA's Quick-Action Superfund Plan Receives Flak from Commenters." *Durango Herald*, September 14. Print.

Rosemeyer, Tom. 2017. "History and Mineralogy of the Sunnyside Mine, Eureka Mining District, San Juan County, Colorado." http://geoinfo.nHmt.edu/museum/minsymp/abstracts/view.cfml?aid=100.

"Technical Evaluation of the Gold King Mine Incident, San Juan County, Colorado." 2015. Washington, DC: US Bureau of Reclamation. https://www.usbr.gov/docs/goldkingminereport.pdf.

Thompson, Jonathan. 2015. "Gold King Mine Water Was Headed for the Animas, Anyway." *High Country News*. https://www.hcn.org/articles/acid-mine-drainage-explainer-animas-pollution-epa-gold-king.

US DOI (US Department of the Interior). 2015. *Technical Evaluation of the Gold King Mine Incident*. Denver: US Department of the Interior, Bureau of Reclamation Technical Service Center.

US EPA (US Environmental Protection Agency). 2015. "Gold King Mine." Memorandum 1547102. US Environmental Protection Agency Region 8 to Site File, August 17.

US EPA (US Environmental Protection Agency). 2016. "Gold King Mine Watershed Fact Sheet." http://www.epa.gov/goldkingmine/gold-king-mine-watershed-fact-sheet.

US EPA (US Environmental Protection Agency). 2019. "Final Interim Record of Decision for Bonita Peak Mining District Superfund Site: Operable Unit 1, San Juan County, Colorado." Denver: US EPA Region 8 Headquarters.

US House Committee on Natural Resources. 2016. *EPA, the Department of the Interior, and the Gold King Mine Disaster*. Majority staff report, February 11. Washington, DC: US House of Representatives, 114th Congress.

1

A Tale of Two Places

The Upper and Lower Animas River Watersheds in Southwest Colorado

BRAD T. CLARK

This chapter begins with a brief summary of Colorado's history to provide context for later discussions covering the political and socioeconomic characteristics of the Animas River Basin. The narrative is organized in two parts, around two distinct watersheds. The first involves the Upper watershed, where the Animas begins as a trickle in a high-altitude, alpine setting in San Juan County, Colorado. It flows by mining-era ghost towns from the mid- to late 1800s and scores of abandoned gold and silver mills and piles of mine tailings. Small tributaries also drain from alpine basins, where mine shafts and tunnels were blasted into mountainsides at incredible elevations. These now abandoned sites are the sources of today's acid mine drainage (AMD), which has degraded the environment and severely damaged water quality through pollution from heavy metals.

When the Animas reaches the county seat at the Town of Silverton, it is joined by Cement Creek, which drains the once lucrative Eureka Mining District where the abandoned Gold King Mine (GKM) is located. After leaving Silverton, the river quickly descends through the rugged Animas River

DOI: 10.5876/9781646421756.c001

canyon and crosses into La Plata County, Colorado, where its gradient and velocity slowly lessen.

This section of the Animas is the subject of the second part of this chapter. The Lower watershed is defined from the point where the Animas flows beneath the historic Baker's Bridge north of Durango, Colorado, to the New Mexico border. At Baker's Bridge the river emerges from the confines of the Animas River canyon and meanders calmly through farmlands and ranchlands. At Durango, the county seat of La Plata County, the river's pace increases, tumbling over natural and human-made whitewater rapids. This is an urban environment, with a paved path paralleling the Animas; it is heavily used by cyclists and pedestrians, and the river hosts boaters and anglers. After leaving Durango, the Animas continues southeast, through irrigated fields and pastures, before crossing into the Southern Ute Reservation and the State of New Mexico, where it joins the San Juan River near the City of Farmington before entering the Navajo Reservation and ultimately crossing into Utah.

PART ONE

The State

The first formal organization of lands that would eventually become the State of Colorado occurred in 1859, when the Territory of Jefferson was established. It was roughly 70 percent larger than the modern state, covering parts of the Kansas, Nebraska, New Mexico, Utah, and Washington Territories. Jefferson was not officially recognized by the federal government and was independently governed by a democratically elected Provisional Government until 1861, when the US Territory of Colorado was formally established. Colorado existed as an organized territory of the United States until 1876, when on August 1—28 days after the nation's centennial—President Ulysses S. Grant signed a proclamation to admit the State of Colorado to the Union. Hence its moniker, the "Centennial State."

Human habitation predated statehood by at least 13,000 years (Dell' Amore 2009). Long before European exploration, various Native American cultures and tribes occupied certain areas, generally including the Folsom and Clovis Cultures in the northern and eastern plains; early Puebloan Peoples on the Colorado Plateau; the Apache, Comanche, and Cheyenne on the eastern plains; the Arapaho along lower elevations of the Front Range; and the Shoshone along the intermountain valleys in the northern parts of the state. In addition, various bands of Utes controlled large areas west of the Continental Divide.

Spanish conquistadors arrived in southern and eastern portions of the state during the 1600 and 1700s. Centuries later, fortune seekers heading to California found gold in the early 1850s along creeks in an area that would ultimately become the Denver metro area. Subsequent discoveries of gold continued throughout the decade, with many of the successful miners (so-called Fifty-Niners) settling the early Denver area (Brace and Virga 2009, 37–50). As the years progressed, prospectors followed the Colorado Mineral Belt to the south and west, eventually penetrating the rugged San Juan Mountains.[1] By the early 1870s, they found promising deposits of precious metals and minerals; the San Juan mining boom had officially begun (Noel 2015, 71).

The Animas River

Located in Colorado's rugged southwest corner, the river received its name in 1765 when Spanish explorer Juan Maria Antonio de Rivera referred to it as el Rio de las Animas.[2] The river begins near the mining-era ghost town of Animas Forks, where the west and north forks of the namesake river meet at an elevation of over 11,100 feet. It first flows west and then in a southerly direction for roughly 126 miles, through the communities of Silverton and Durango, Colorado, and Aztec, New Mexico, until its confluence with the San Juan River near an elevation of just under 5,400 feet. In all, the Animas River drops more than 5,700 feet from its high alpine headwaters to the semi-arid deserts of northern New Mexico. In terms of area, the Animas River's drainage basin size ranks 30th (out of 74) in Colorado (Gustafson 2003).

The long-term annual average flow of the Animas at Silverton is approximately 125 cubic feet per second (CFS). Because of snowmelt and summer monsoon storms, the river is usually at its highest in June and July and lowest in January and February. The highest annual mean flow measured at Silverton was 194 CFS in 1995; the lowest was 53 CFS in 2002 (US Geological Survey 2017).

When gauged at Durango, the long-term annual average flow of the Animas is approximately 809 CFS. The highest annual mean flow in Durango was measured in 1917 at 1,366 CFS. The lowest was in 2002 at 238 CFS (US Geological Survey 2017).

In New Mexico, the annual average flow of the Animas is approximately 904 CFS, as measured just south of the Colorado border at Cedar Hill. The highest annual mean flow was 1,713 CFS in 1941. The lowest was recorded at 268 CFS during the 2002 drought (US Geological Survey 2017).

A TALE OF TWO PLACES: THE UPPER AND LOWER ANIMAS RIVER WATERSHEDS

Although there is no official point of demarcation between the Upper and Lower Animas River watersheds, it is generally understood that the landmark Baker's Bridge provides an acceptable dividing line between the two.[3] This claim is based on several changes in the river's character below this spot, including a notable lessening in gradient and velocity, improvements in water quality, numerous agricultural diversions and irrigated farming and ranching, and changes in land ownership (i.e., from public to private).[4] With Baker's Bridge as the dividing line, the mainstem Upper Animas River flows roughly 45 miles from its source near Animas Forks.

The Upper Animas: San Juan County and Silverton, Colorado

From its beginning, the Animas River flows through the mining-era ghost towns of Eureka and Howardsville before reaching Silverton. At Silverton, two tributaries—first Cement Creek, then Mineral Creek—enter the Animas; both contain high levels of sulfuric acid and dissolved heavy metals, which severely hinder water quality and aquatic life. Although some of this is natural due to the region's heavily mineralized geology, hundreds of hardrock mines were dug into the sides of these two tributaries, creating AMD-related portals and exposing large amounts of mineral-laden mine tailings. As tributary flows increase over the next 30-odd miles, water quality generally improves as the Animas flows through the narrow depths of the Animas River canyon.

Today, whitewater rafting and kayaking are the main recreational activities in this stretch of the Animas, where Class IV and V rapids are the norm. This is particularly the case for the roughly 27 miles between Silverton and the Railyard Access at Rockwood, which is widely considered one of the longest

Class IV–V whitewater runs in the United States. Boating on the few remaining miles of Class V+ rapids between Rockwood and Baker's Bridge, known locally as Baker's or Pandora's Box, is unadvised due to the river's containment in a vertical box canyon, with little to no chance for rescue or escape. In addition, many of the near-continuous rapids and small pour-overs are un-scoutable and un-portagable.

Another major, albeit calmer, tourism-related activity is the Durango & Silverton Narrow Gauge Railway (D&SNGR), which generally parallels the Upper Animas on its 45 mile run between the two namesake towns. The rail line has operated continuously since its completion in 1882; during its first four decades, the D&SNGR was primarily a freight line, transporting supplies to Silverton-area miners and mineral-rich ores to Durango. Since the early 1950s, the D&SNGR has operated solely as a passenger train, currently with 200,000 annual customers.[5] The tracks south of Durango were abandoned in 1969, thus cutting it from New Mexico and isolating the line exclusively between Durango and Silverton.

The Animas's headwaters are located in San Juan County, which was established 3 months prior to statehood on January 31, 1876. Due to its remote and difficult-to-reach location, it was the last county established in the Colorado Territory. San Juan County's name came from its location in the San Juan Mountains and near the headwaters of the San Juan River's largest tributary, the Animas. Both of these geographic features were named previously by eighteenth-century Spanish explorers in honor of Saint John the Evangelist, more commonly known as John the Apostle.

In terms of contemporary land ownership, 88.5 percent of San Juan County is federally owned, with approximately 71 percent administered by the US Forest Service and 18 percent under US Bureau of Land Management jurisdiction. State of Colorado lands account for roughly 0.50 percent. Privately owned lands account for the remaining 11 percent and consist mostly of patented mining claims. The result is a checkerboard-type pattern of land ownership, with small inholdings of private lands scattered among public lands. An estimated 2,213 mining claims are currently owned in San Juan County; the vast majority are private (as opposed to corporate) and only ten mines are considered active ("Mining Companies" 2017).

In terms of physical size, San Juan County ranks as the state's fifth smallest, with an area of 387,40 square miles. Initially, the county was much larger than it is today, with its original boundaries reaching from the Utah border in the west to its present-day border in the east. Its current boundaries became fixed in 1877, when most of the county's western portion was reallocated to form Ouray County during the first meeting of the Colorado General Assembly.

Because of its remote alpine setting, abundance of natural resources, and rich history of hardrock mining, San Juan County is home to all or parts of ten nationally protected areas. Among these are three national forests (Rio Grande, San Juan, and Uncompahgre), four national historic districts (Durango & Silverton Narrow Gauge Railway, Silverton, Martin Mining Complex, and Tabasco Mine and Mill), two national historic landmarks (Shenandoah Dives / Mayflower Mill and Cascade Boy Scout Camp), and one wilderness area (Weminuche). In addition, two national (automotive) byways—the Alpine Loop Backcountry and the San Juan Scenic Skyway—travel through the county, as do two nationally renowned hiking routes—the Colorado Trail and the Continental Divide National Scenic Trail. Twelve of the state's fifty-four peaks above 14,000 feet rise above San Juan County, and it holds the distinction of being the nation's highest county, with an average elevation of 11,240 feet above sea level. Together, in 2016 these attractions drew an estimated $17.8 million in travel-related spending and $700,000 in local taxes to the Silverton area, making tourism its modern economic mainstay (Dean Runyon Associates 2016). Of the five counties in the southwest Colorado Development District, tourism is undoubtedly the most important industry in San Juan County, providing more than 40 percent of all jobs ("Tourism as a Base Industry" 2015).

Over a century after the first Spanish explorers entered the San Juan Mountains, hardrock miners came to the area of present-day Silverton and San Juan County. They encountered Native American bands associated with the Ute Tribe and culture, which had occupied most of the region for centuries. Despite attacks and attempted encroachments by the Arapaho and Cheyenne Tribes, the Utes controlled most of present-day Colorado that is located west of the Continental Divide. Ute control began to unravel after the prospector Charles Baker—widely regarded as the first white American to enter the San Juan County area—arrived in 1860 (Ninnemann and Smith 2006). Baker and members of his party found gold and silver deposits along the Animas River in an area later known as Baker's Park. Exaggerated news of the find led to the arrival of hundreds of miners the following year.

During the US Civil War, mining in San Juan County decreased substantially, yet by 1870 mining activity had rebounded as entrepreneurial miners located a main source of the gold and silver deposits first found by Baker and his party—the Little Giant Mine. By 1872, silver and gold ore worth roughly $30,000 per ton was being extracted from lands that would become San Juan County (Colorado Encyclopedia 2017).

Momentum to officially establish the county peaked in 1873, when the Brunot Agreement (also known as the San Juan Cession) was negotiated between the United States and the Ute leader, Chief Ouray. After more than four centuries of Ute occupation, the United States took control of 4 million acres in the San Juan Mountains. In exchange, the Utes were paid seven-and-a-half cents per acre and promised control of the region's river valleys, which were highly coveted for their agricultural potential (Colorado Encyclopedia 2017). Utes' command over their river valleys never materialized, and they were effectively dispossessed from their homeland and relocated along with other area tribes to the Uintah and Ute Reservation, which had been established previously in 1864 in northeastern Utah. Descendants of the Mouache and Capote bands of Utes were moved to the Southern Ute Reservation (1873) located in southwestern Colorado, while Ute descendants of the Weminuche band were relocated to the Ute Mountain Ute Reservation (1897), located in southwestern Colorado, northwestern New Mexico, and an isolated part southeastern Utah (i.e., White Mesa).

By the mid-1880s, San Juan County was home to more than 100 mines, which generated a combined total of roughly $750,000 in silver, $207,000 in lead, and $40,000 in gold (Colorado Encyclopedia 2017). One area in particular, the Eureka Mining District, was home to two of the county's most productive mines—the Sunnyside Mine (patented 1874) and the Gold King Mine (GKM) (patented 1887). These two Eureka District mines were established within the 10 square mile collapsed Silverton caldera—the richest geological area in the entire San Juan mining region (Burbank and Luedke 2017).

The GKM was first discovered by prospector Olaf Nelson in 1887, near the headwaters of the Cement Creek drainage. It did not become fully operational until 1893, two years after Nelson's death. GKM was sold the following year for $15,000; by 1895, significant amounts of valuable ore were being produced at GKM and a number of other shuttered mines that had been reopened. By 1896, San Juan County mines produced an estimated $1.5 million in silver, $909,000 in gold, and $169,000 in lead—a remarkable feat given that the nation's economy was limping through a series of recessions and market panics during this same time. In 1902, British investors offered to buy GKM for $4 million, but the new owners declined (Brasch 2015). Eventually, GKM grew to include forty claims and a large stamp mill at the nearby settlement of Gladstone. Daily output at GKM would grow to an estimated 400 tons of ore, which yielded 50–60 tons of high-value concentrates in gold, silver, and copper (Brasch 2015).

Although it was subject to market fluctuations and the boom-and-bust cycles typical of mining areas, hardrock mining remained the most important part of the San Juan County economy through the beginning of the twentieth century. By the end of World War I, county mines had produced more than $60 million in gold, silver, copper, lead, and zinc. Full-scale operations at GKM ended in 1923. The Great Depression and, later, World War II brought an effective end to large-scale mining in the region. In 1991, the last major mine in the Silverton area closed.

According to the US Census Bureau, San Juan County had an estimated population of 694 residents in 2016, making it the smallest of Colorado's 64 counties and among the 100 least populous in the United States. The resident population of San Juan County is expected to maintain through 2035 ("Economic Snapshot" 2017). The Town of Silverton is the county seat, with an estimated 2016 population of 630 (US Census Bureau 2017c). It is the only incorporated municipality in the county, which lends to its distinction as a "one town county." These population figures do not include the large number of seasonal visitors, many of whom own second homes in the county.

In terms of income, San Juan County had a 2015 median household income of $36,324 (US Department of Commerce 2017). In 2016, workers earned an estimated $504 in weekly wages, which is less than half the national average and the lowest of all Colorado counties (US Department of Labor 2017). The 2015 unemployment rate was an estimated 4.5 percent, slightly higher than the Colorado average ("Region 9 Economic Quarterly" 2017). This figure is remarkable, given that unemployment reached an all-time high of 78.2 percent in 1991, the same year the last remaining large-scale mine in San Juan County closed. An estimated average of 16.5 percent of San Juan County residents were considered as living below the poverty level between 2011 and 2015 (US Department of Commerce 2017).

The hardrock mining boom and subsequent growth of the Silverton area occurred at a time in American political history that political science scholars commonly refer to as the era of personal or celebrity politics, as politicians were known more for their popular personas and styles of communication than for their partisan stances and policy preferences. The two most prolific national political figures during this era were Theodore Roosevelt and his Democratic opponent, William Jennings Bryan.

Jennings Bryan was known as the "Great Commoner" because of his steadfast and romantic espousal of common Americans, especially in terms of their inherent wisdom, faith, and hard work (Kazin 2006, 88). Because of this, Jennings Bryan was especially popular among so-called producing classes of Americans, which included farmers, workers, miners, and generally rural as opposed to urban populations (Postel 2007). Unlike many other populations in the American West, particularly those without histories of silver mining, residents of San Juan County had a particular affinity for William Jennings Bryan and his Populist Democratic contemporaries.

In particular, Jennings Bryan supported the concept of bimetallism, the economic principle behind a monetary system whereby the worth of the monetary unit is seen as equivalent to a fixed quantity of and exchange rate between two metals—usually gold and silver (Velde and Weber 2017). This support for two metals, as opposed to the gold standard alone, made him especially popular among a short-lived faction of the Republican Party—the Silver Republicans, who were prominent in silver mining regions in the American West where "free silver" was a popular economic policy. It favored unrestricted coinage of silver as money on demand, as opposed to strict reliance on the more rigidly fixed money supply inherent with the gold standard (Painter 2008). Simply put, farm and other labor groups believed that with an increased amount of currency in circulation (i.e., gold *and* silver), their primary commodities would receive higher prices.

This combination of support for working-class Americans and his endorsement of the free silver policy provided William Jennings Bryan with strong support from hardrock mining communities such as Silverton. To a notable degree, this explains why San Juan County voted Democratic in presidential elections between 1892 and 1916 and continued to lean toward the Democratic Party until the early 1940s, when Republican candidates began to make inroads (Leip 2017).

Throughout the last third of the twentieth century, San Juan County leaned in a decidedly Republican direction in national elections. This partisan shift paralleled the demise of large-scale hardrock mining in the southern Rockies. In the nine presidential elections between 1968 and 2000, Silverton-area voters favored Republican candidates. The only exception was in 1992, when San Juan County voters awarded the most votes to the Independent candidate, Ross Perot, who tapped into a Populist anti-establishment sentiment that was prevalent in rural, working-class portions of the country such as Silverton. From 2004 through 2016, voters in San Juan County expressed support in the other direction by awarding the most votes to Democratic candidates in the four consecutive presidential elections during that period.

On the surface, it appears that Silverton-area voters have come to wholeheartedly endorse the Democratic Party's platform in recent decades. Yet closer inspection of votes cast in San Juan County, compared to national averages, reveals a more nuanced picture. Specifically, in all seven presidential elections from 1992 through 2016, San Juan County exceeded the national average in support of third-party candidates (i.e., those not affiliated with either of the two main parties), including those representing the Libertarian (2012, 2016), Independent (1992, 2004, 2008), Reform (1996), and Green (2000) Parties. Prior to this, San Juan County voters far exceeded the national average in support of third-party presidential candidates in eleven of the eighteen elections between 1912 and 1980. In 2016, San Juan County had the largest percentage of unaffiliated registered voters (45.3%) in all of Colorado's sixty-four counties (Hamm 2016).

Findings from both long-term studies of American presidential elections and short-term case studies of individual campaigns suggest a number of explanations behind third-party voting behavior (e.g., "American National Election Studies" 2017; Johnston et al. 2004; Rockler 2013). Specifically, Independent-leaning third-party voters tend to generally dislike or distrust government, are more likely to feel disenfranchised by the political process, and tend to view their vote as one against the two-party establishment.

Distrust of government, especially at the federal level and of regulatory agencies like the EPA, has been a defining feature of the political culture unique among Silverton-area residents ("About Us" 2016; Coughlin et al. 1999; Pendley 2017). This combined with fears of losing local control regarding mine reclamation has been a major source of opposition to Superfund designation.

The voting behavior of San Juan County residents in national elections has generally mirrored that of the rest of Colorado. In 2000, the county and the state went for the successful Republican ticket of George W. Bush and Dick Cheney and, along with voters from other counties in District 3, supported the successful Republican candidate, Scott McGinnis, for a seat in the US House of Representatives. In 2002, the same pattern of unified support of Republican candidates emerged in national elections for seats in both the US House (again McGinnis) and the US Senate (incumbent Wayne Allard). Yet similar to the partisan shift that occurred in Colorado during the first decade of the twenty-first century, San Juan County voted for Democratic candidates in the 2004 presidential (challenger John Kerry), Senate (Kenneth Salazar), and House (younger brother John Salazar) elections. This pattern continued in all midterm and presidential elections through 2018, except for the 2014 election in the US House, when San Juan County voters in District 3 supported the successful Republican candidate, Scott Tipton.

Regarding state-level elections for seats in the bicameral Colorado General Assembly, voting behavior among San Juan County residents has exhibited general support of Democratic candidates. San Juan County voters are part of the state's 6th Senate District, where they supported Democratic candidates in all but one (2014) election from 2000 through 2018. In the Colorado House, Republican candidates from District 59 have fared somewhat better among San Juan County voters, receiving the most votes in four (2000, 2002, 2004, and 2008) of the last ten elections (2000–2018).

Eight elected officials are charged with implementing state law and managing business in San Juan County: the assessor, clerk and recorder, coroner, sheriff, and treasurer and the three-member County Board of Commissioners—who represent the chief legislative / executive body of the county. Each commissioner represents one of three districts; together they oversee the annual budget, hire staff, oversee land-use planning and development, and administer county social services.[6]

When the GKM spill occurred in 2015 and during the months leading to official Superfund designation in 2016, the County Board of Commissioners consisted of one unaffiliated and two Democratic commissioners. All three were reelected following Superfund listing. District 1 was represented by

Peter McKay (D). He was first elected in 2000 and was reelected to five consecutive terms. During the final year of his four-year term, 2020, McKay was elected board chair; he indicated that he wouldn't seek a sixth term. District 2 was represented by Scott Fetchenhier (D). He was first elected in 2012 and was reelected to two consecutive terms. Fetchenhier previously served on the Silverton District #1 School Board. District 3 was represented by Ernest Kuhlman (U), who served as chair of the board when Superfund came to the county in 2016. He had over thirty years of public service in elected positions, including Silverton's mayor.

The Town of Silverton Board of Trustees is composed of six trustees and a mayor, all of whom are elected on the first Tuesday in April in even-numbered years. When the GKM spill occurred in 2015, the town's board consisted of two Republicans, two Democrats (including the mayor), and three unaffiliated trustees. When official Superfund designation was announced after a unanimous vote by the board in 2016, the partisan makeup had shifted to three Republicans, one Democrat (the mayor), one Green Party member, and two unaffiliated trustees.

As of November 2017, three of the trustees were registered Republicans, two were unaffiliated, and one was registered with the Green Party. The mayor was a registered Democrat. All terms are for four years, and office-holders are limited to two consecutive terms. Together, the trustees serve as the town's Finance Committee, while individual trustees act as chairpersons for the town's remaining standing committees—Buildings and Grounds; Public Safety; Utilities; Code and Personnel Manual; and Parks, Recreation, and Events. The total general fund revenue for Silverton's 2017 adopted budget was roughly $2.3 million ("Budgets / Audits" 2017).

The town government also has multiple administrative departments to handle daily operations in areas such as finance, planning, building and fire inspection, public works, and code enforcement. The town administrator, who is appointed by and serves at the discretion of the trustees, oversees these departments ("Government" 2017b). The town's Public Works Department is of particular relevance to this work, as it provides drinking water for Silverton residents. It is sourced from two drainages (Boulder and Bear Creeks) located just outside the highly mineralized Silverton caldera, where mining activities have been prohibited since the nineteenth century (Frodeman 2003, 39).

County commissioners and town trustees come together via the San Juan Regional Planning Commission and work to develop a master plan to guide physical growth in the town and the county. They decide on zoning and development issues and make recommendations regarding the approval of subdivisions and plat amendments ("Government" 2017a).

Following the 1860 discovery of gold and silver deposits by the Baker Party, a hardrock mining boom was imminent. The first rush occurred in 1874, when an estimated 2,000 prospectors came to the Baker's Park area of the valley, where they established around 1,000 mining claims (National Park Service 2008). The area quickly became the town site of Silverton, which was incorporated in November 1876. As the established hub of mining activity in the Upper Animas watershed, Silverton reportedly was named in response to a comment expressed by a local prospector: "We do not have much gold but we have silver by the ton" (Evans 2010).

To facilitate commerce both to and from Silverton, a toll road to the south was constructed between 1876 and 1877. The Animas Canyon Toll Road connected the lower Animas River Valley with Silverton as a means to both import food supplies and material resources to Silverton and export precious metals and mineral ores to Animas City (and later Durango). The roughly 32 mile road cost an estimated $32,000 to complete; it was crucial for opening the upper Animas Valley to intensive settlement, as teams of mules and wagons imported endless supplies of coal for fuel, fresh produce and meat for hungry miners, and hay for livestock (National Park Service 2008).

In July 1881, the Denver & Rio Grande Railroad reached the modern-day location of Durango; a year later, the line was extended northward along the Animas River to Silverton. This rail linkage provided the foundation on which both communities grew in the remaining decades of the nineteenth century. Mining activity and the wealth it produced led to construction of urban infrastructure, a large smelter, stamp mills, and a string of saloons and brothels along Silverton's infamous Blair Street.

By 1896, the majority of Colorado mining regions were failing or had become mired in economic depression, yet several factors allowed San Juan County to buck this trend and actually make significant gains. First, the remaining mines, especially the largest ones, were still producing rich gold and silver ore necessary to keep local and regional smelters in business. Second, wealthy investors such as Otto Mears—builder of the Animas

Canyon Toll Road—injected significant capital into mine expansion, road maintenance, and railroad construction. The result was an increase in mine productivity (Colorado Encyclopedia 2017).

Despite local labor disputes and political developments at the national level, hardrock mining (especially for silver) remained the most important driver of San Juan County's economy through the beginning of the twentieth century. For example, in 1891, the region produced more than $761,000 in silver compared to $192,000 in gold (Blair 1996). In fact, an overproduction of silver in areas such as San Juan County led the federal government to pass the Sherman Silver Act of 1890, which required that the government purchase a large amount of silver on a monthly basis to stimulate demand for surpluses of the precious metal. At first, this artificially over-inflated the price of silver relative to gold, which led to a serious decline in the amount of gold in circulation. This occurred because under the 1890 act, the government paid sellers of silver via specially printed US Treasury notes in specific denominations, which could then be redeemed in private metals markets for coinage of either silver *or* gold—where gold prices exceeded the government's exchange rate for silver. Thus an increasing number of people sold silver to the federal government at the artificially inflated rate, exchanged their Treasury notes for gold coins, and proceeded to sell them in the metals market for prices higher than they had initially paid for the silver. Profits from such transactions were then used to purchase more silver and repeat the process; ultimately, the federal Treasury's supply of gold reached precariously low levels.

In large part, this led to the so-called Panic of 1893, which prompted congressional repeal of the Sherman Silver Act and the resultant collapse of silver prices. In 1890, silver was valued near $1.20 per ounce; by the end of 1894, it had fallen to below $0.60 an ounce. The effects were disastrous for mining regions across the West, as countless silver mines closed. According to Colorado historian Duane Smith (1996), the era in which individual miners staked their claim and made a fortune was henceforth over; mining in San Juan County was taken over by large-scale corporate enterprises, whose financial reserves allowed area mines to survive the Panic and actually increase production in ensuing years. By the end of World War I, Silverton-area mines had produced more than $60 million in gold, silver, copper, lead, and zinc (Smith 1996). By 1948, this figure had grown to more than

$123 million ("Geology and Ore Deposits" 2017). After World War II, however, mining in San Juan County had largely come to an end.

By the late 1950s, the main commodity hauled by the D&SNGR between Durango and Silverton shifted to tourists and, later, outdoor recreationalists. In recent decades, San Juan County has become almost entirely dependent on tourism, primarily during the summer months when the D&SNGR is in full operation. A year following the 2015 GKM spill, ridership on the tourist train fell to its lowest August level since at least 2002 ("Tourism Data" 2017).

Notwithstanding, winter tourism has grown since the 2002 opening of the Silverton Mountain ski area—currently the county's largest single employer ("Comprehensive Economic Development Strategy" 2016). In addition to the ski area, top employers in San Juan County in 2014 included Handlebars Restaurant (n = 26), San Juan County (n = 26), Brown Bear Café (n = 22), Town of Silverton (n = 22), Silverton School District #1 (n = 12), and the Pickle Barrel Restaurant (n = 10) ("Comprehensive Economic Development Strategy" 2016). A 2014 analysis by the Colorado Demography Office reaffirmed that tourism is the largest countywide economic driver, providing 50 percent (or 154) of base industry jobs (i.e., those that infuse income from outside the community) ("Town of Silverton" 2016). In 2016, employment in service-sector industries accounted for 51 percent of all jobs in San Juan County, a figure nearly ten points higher than that for La Plata County and five points higher than the statewide average of 46 percent ("Region 9 Employment Data Update" 2016).

The Lower Animas: La Plata County and Durango, Colorado

The distance from Baker's Bridge to the next river access point at Durango's northern end is roughly 20 miles of meandering flatwater and oxbows through mostly private agricultural land. The gradient increases for the Animas as it flows through the City of Durango. This roughly 10 mile urban stretch is very popular during the late spring and summer months for rafting, tubing, and fishing. After leaving Durango, the river travels nearly 20 miles to the New Mexico state line; most of this stretch flows through the Southern Ute Reservation. The Animas continues for approximately 30 additional miles to its confluence with the San Juan River, just south of Farmington, New Mexico. In total, the Lower Animas River—from Baker's Bridge to the San Juan confluence—is approximately 81 miles.

La Plata County was established on February 10, 1874, roughly two-and-a-half years prior to Colorado's statehood. Its name was a reference to the La Plata River and La Plata Mountains located just west of Durango. Both were named previously by eighteenth-century Spanish explores because of the large silver deposits found in the vicinity of these two geographic features; the Spanish word for silver is *Plata*. The City of Durango was named in reference to Durango, Mexico; the name comes from the Basque word for "water town"—*urango*.

In terms of contemporary land ownership, roughly 39 percent of La Plata County is federally owned, with approximately 37 percent administered by the US Forest Service and 2 percent under US Bureau of Land Management jurisdiction. State of Colorado lands account for roughly 2 percent, while privately owned lands in the county constitute roughly 43 percent. La Plata County is also home to two Native American reservations; the Southern Ute Tribe occupies approximately 17 percent, while the Ute Mountain Ute tribal lands cover just over 0.001 percent of the county. These large percentages of private and tribal lands, a total of roughly 50 percent, are in stark contrast to those in San Juan County, which has no reservation lands and only 11 percent of its land in private ownership (Congressional Research Service 2020).

In terms of physical size, La Plata County ranks as the state's 38th largest county, with an area of 1,700.44 square miles. Its northern border is defined by San Juan County and its southern border is the New Mexico state line. An estimated 6,010 mining claims are currently owned in La Plata County; the vast majority are private (as opposed to corporate), and only thirty-seven mines are considered active ("Mining Companies" 2017).

Geographically, La Plata County is more varied than its neighbor to north. While San Juan County is virtually all high alpine environment, La Plata County covers a region with landscapes ranging from 13,000+ foot alpine peaks and alpine meadows in the north to high desert plateaus and mesas in the south. It is also home to a significant portion of both the San Juan National Forest and the Weminuche Wilderness Area and is the sole home of the Hermosa Creek Wilderness Area, established in 2014. Related to its Native American, Spanish, and mining histories, La Plata County is home to six districts and eight landmarks of archaeological, historical, and industrial importance listed on the National Register of Historic Places (2017). In addition, one national byway—the San Juan Scenic Skyway—and three national trails—the Colorado Trail, the Old Spanish National Historic Trail, and the Great Parks Bicycle Route—travel through the county. Together, these and other tourism-related attractions provided nearly 20 percent of all jobs in La Plata County in 2016, confirming that tourism is among the area's modern economic mainstays ("Regional Data" 2017).

Dating as far back as 6000 BC, Paleo-Indian peoples led nomadic lives in areas around today's Four Corners region. As conditions allowed for more permanent settlement and limited agriculture, one pre-Ancestral Puebloan culture known today as the Basketmaker became well established by 500 BC across areas of the Colorado Plateau (Cassells 1997). By the mid-AD 700s, Ancestral Pueblo peoples occupied areas of La Plata County, including sites near modern-day Durango as well as in greater concentrations associated with the Mesa Verde complex. They were all connected to the broader pattern of Ancestral Puebloan settlement, ranging from southeastern Utah to southwestern Colorado. Similar to what happened at Mesa Verde in the AD 1300s, these early inhabitants were part of a massive southward migration into parts of present-day Arizona and New Mexico.

By the early 1300s, bands of Utes had migrated to the Four Corners region from the west. By the 1600s, roughly a half-dozen loosely connected Ute bands occupied parts of what would become Colorado; two of them, the Mouache and Capote, settled in areas of present-day La Plata County. The earliest known European contact occurred in 1765, when Spanish explorers led by Juan Maria Antonio de Rivera entered the region. This was followed by the Dominquez and Escalante expeditions in the mid-1770s and later by travelers on the Old Spanish Trail that connected Santa Fe, New Mexico, with areas in Southern California.

As was the case with the Weminuche-affiliated Utes in the Upper Animas watershed, Ute control in the Lower watershed steadily eroded as increasing numbers of Anglo-Americans followed the California and Colorado Gold Rushes. The 1873 Brunot Agreement and 1887 Dawes Act, which divided Native American lands into individual allotments based on tribal affiliation, effectively ended Ute occupation and control of the region. The hardrock mining activity that had descended onto San Juan County spread south to La Plata County and ultimately led to the establishment of a small settlement by Charles Baker in the winter of 1860. First called Animas City, the settlement was subsumed by the emergent regional railway hub of Durango by 1948.

Modern-day Durango owes its existence in large part to the decision made by the Civil War hero and railroad engineer General William Jackson Palmer to extend his Denver & Rio Grande Railroad westward from Chama, New Mexico, along the Animas River to Silverton, Colorado. Assuming that the new rail line would pass through their settlement, the 2,000+ homesteaders who had incorporated Animas City by 1878 inflated land prices in anticipation of financial rewards (Smith 2007). After failing to negotiate agreeable terms with the Animas City residents, General Palmer staked out his own town, Durango, a few miles to the south. By 1880, Durango had outsized Animas City and was officially incorporated on September 13, 1880 (Smith 2007).

After the rail tracks from Durango to Silverton were completed in July 1882, mining operations in the Upper watershed were bolstered, and Durango became the main distribution point to supply the entire San Juan Mining District (Smith 2007). General Palmer also oversaw construction of the New York and San Juan Smelter (NYSJS) in Durango in 1882 to process the gold, silver, lead, and copper ores extracted from higher elevations. The nearby discovery of large coal deposits allowed it to become the largest smelting operation in the Four Corners region, and Durango soon emerged as a leading economic hub in the southwestern part of the state. To support its burgeoning population, agricultural operations expanded throughout the Animas River Valley, and ranches in the mountains and plateaus to the west of Durango produced increasingly large numbers of cattle and sheep ("Permanent Settlement" 2008). By the mid-1920s, 360,000 acres had been put into cultivation by La Plata County farmers (Seyfarth and Lambert 2010).

The near collapse of hardrock mining in San Juan and La Plata Counties around the time of the Great Depression led to the closure of the NYSJS by 1930. The US Vanadium Corporation replaced it in the early 1940s with a uranium processing mill where the US military produced uranium for the Manhattan Project and, ultimately, for the atomic bombs dropped on Hiroshima and Nagasaki. Uranium milling continued at the site until 1963; in 1985, the US Department of Energy approved comprehensive remediation at the site, along with hundreds of others in and around Durango. Contaminated soils were buried west of the city (Smith 2007).

In the 1970s, La Plata County's economy began to transition from a traditional reliance on agriculture and extractive industries such as mining to jobs in the service sector, including business, professional, and health services;

education; and tourism ("Economic Snapshot" 2017). Fossil fuel development in La Plata County, particularly natural gas, began to decline notably by early 2013 and reached an all-time low of 66.3 million cubic feet in April 2016 ("Oil and Gas" 2017). Prior to this, by 2015, employment in the service sector accounted for 42 percent of all jobs and 35 percent of earnings in La Plata County ("Economic Snapshot" 2017). Top employers in this sector in 2016 included Purgatory Resort (n = 1,000), Mercy Regional Medical Center (n = 860), Durango School District 9R (n = 853), Southern Ute Tribe (n = 795), Fort Lewis College (n = 660), and Sky Ute Casino (n = 425) ("Economic Snapshot" 2017).

According to the US Census Bureau (2017b), La Plata County had an estimated population of 55,623 residents in 2016, making it the 50th largest of Colorado's sixty-four counties. The City of Durango is the county seat, with an estimated 2016 population of 18,503 (US Census Bureau 2017a). It is the county's most populous municipality. Both populations are expected to grow at moderate rates through 2040. These population figures do not reflect the large number of seasonal visitors, many of whom own second homes in the county.

In terms of income, La Plata County had a median household income of $60,278 in 2015 (US Department of Commerce 2017). The unemployment rate for the same year was an estimated 3.9 percent, a figure slightly below the Colorado average of 4.2 percent ("Region 9 Economic Quarterly" 2017). An estimated average of 10.6 percent of La Plata County residents were considered as living below the poverty level between 2011 and 2015 (US Department of Commerce 2017). In 2015, any figure above $27,012 defined a livable wage for a single person renting a one-bedroom apartment. In 2016, workers earned an estimated $935 in weekly wages, which is slightly less than half the national average of $1,067 ("County Employment" 2017).

Similar to that previously discussed for San Juan County, voting behavior of La Plata County residents in national elections has generally mirrored that of the rest of Colorado. In 2000, the county and the state backed the successful Republican ticket of George W. Bush and Dick Cheney and, along with voters from other counties in District 3, supported the successful Republican candidate in the US House election. In 2002, La Plata County also supported the Republican candidate for a seat in the US House but differed from both statewide and San Juan County voters by supporting Tom Strickland, the unsuccessful Democratic challenger to incumbent Wayne Allard (R) in the US Senate. Similar to the broad partisan shift that occurred in Colorado during the early 2000s, La Plata County voted for Democratic candidates in the 2004 presidential, US Senate, and US House elections. This pattern continued for all midterm and presidential elections through 2018, except for the 2014 election to the US House, when La Plata County voters in District 3 (along with San Juan County) supported the successful Republican candidate, Scott Tipton.

Regarding elections to the Colorado General Assembly, voting behavior among La Plata County residents has generally mirrored that of voters in San Juan County, resulting in slightly less support of Democratic candidates. La Plata County voters are also part of the state's 6th Senate District, where they supported Democratic candidates in all but two (2010 and 2014) elections from 2000 through 2018. Similar to San Juan County, Republican candidates from District 59 in the Colorado State House have fared somewhat better than Democrats among La Plata County voters, receiving the most votes in four (2000, 2002, 2004, and 2008) of the last ten elections (2000–2018).

COLORADO ELECTIONS AND GOVERNMENT

Similar to San Juan County residents' voting behavior in presidential elections, voters in La Plata County exceeded the national average in support of third-party candidates in 2012 and 2016 (Libertarian); 1992, 2004, and 2008 (Independent); 2000 (Green); and 1996 (Reform). La Plata County voters also exceeded the national average in support of third-party presidential candidates in fourteen of the eighteen elections between 1912 and 1980. In 2016, nearly half (45.3%) of registered voters in La Plata County were unaffiliated (i.e., not Democrat, Republican, Libertarian, or Green) (Hamm 2016).

The fact that voting behavior in San Juan and La Plata Counties has been nearly identical since 2000, with the exception of the 2002 US Senate race, is curious when considering their markedly different population sizes and demographic profiles—especially in the areas of income, economy, and education. As discussed below, the median household income in La Plata is nearly twice that of San Juan County, as are percentages of residents with a college education (roughly 57% compared to 30%) and graduate degrees (roughly 18% to 10%) ("Colorado" 2017). Furthermore, San Juan County's contemporary economy is almost entirely dependent on tourism, while the economy in La Plata County is more diversified—it includes a regional medical center and an airport, the Southern Ute tribal government, and a four-year public college offering both undergraduate and graduate degrees. Comparatively, the voting behavior in recent national elections among neighboring counties of both San Juan and La Plata Counties has been largely and uniformly different. Specifically, voters in Archuleta, Montezuma, and Montrose Counties have expressed more preference for Republican candidates ("Colorado Elections" 2017). An explanation for the similarities in voting behavior between San Juan and La Plata Counties is thus reduced to the main commonality unique to both counties—their hardrock silver mining histories, linked by the D&SNGR and the Animas River.

Ten elected officials are charged with implementing state law and managing business in La Plata County: the assessor, clerk and recorder, coroner, district attorney, sheriff, surveyor, treasurer, and the three members of the County Board of Commissioners—which represents the chief legislative and executive bodies of the county. Each commissioner represents a specific district, yet they are elected at large; together they oversee the county's budget, oversee land use, and set the county's general direction ("Board of County Commissioners" 2017).

When the GKM spill occurred in 2015 and during the months leading to official Superfund designation in 2016, the county's board consisted of two Democrats and one Republican. District 1 was represented by Brad Blake (R), who was first elected in 2014. District 2 was represented by Gwen Lachelt (D), and District 3 was represented by Julie Westendorff (D). Lachelt and Westendorff were first elected in 2012 and were reelected to second terms in 2016.

All policy and legislative decisions in Durango are made by the five-member City Council. Councilors represent the community at large rather than specific districts or precincts. Elections are nonpartisan. The positions of mayor and mayor pro tem are held by sitting council members and are elected by their fellow counselors to one-year terms. When the GKM spill occurred in 2015, the City Council consisted of four Democrats, including the mayor, and one Republican. When official Superfund designation came in 2016, this partisan composition had changed slightly to include three Democrats, one Republican, and one unaffiliated councilor.

The city government also has multiple administrative departments to handle daily operations in areas such as community development, code enforcement, finance, parks and recreation, police, and transportation. The city manager, who is responsible to the City Council, oversees these departments, as well as the offices and agencies of the municipal government. The total general fund revenue in the 2017 adopted budget for the City of Durango was roughly $37.7 million ("2017 Budget" 2017).

The city's Utilities Department is of particular relevance to the present work, as it provides drinking water for Durango residents. Since the early 1900s, the Florida River has been the city's primary source due to water quality concerns from unregulated dumping of mine waste into the Animas near Silverton. The Florida River watershed is unconnected to the Upper Animas and thus has not historically been impacted by mining activities in San Juan County. The Florida gathers its flow from the Weminuche Wilderness Area northwest of Durango. During summer months, the city's supply may be augmented with water from the Animas; it is assumed that whatever heavy metal contaminants might be introduced will settle to the bottom of the City Reservoir prior to entering the treatment plant.

In 2016, employment in service-sector industries accounted for just over 42 percent of jobs in La Plata County, a figure slightly less than the statewide average of 46 percent ("Region 9 Employment Data Update" 2016). Tourism in particular generated 28 percent of all economic activity in the county. As a primary indicator of tourism strength, the lodgers' tax increased by 7.5 percent over 2015 totals ("Comprehensive Annual Financial Report" 2017). This accounted for a 0.5 percent increase in total jobs in the accommodations and food sector of the county's economy. The strongest area of job growth in La Plata County during this same period was in arts, entertainment, and recreation, with a 10 percent increase between 2015 and 2016 ("Region 9 Employment Data Update" 2016). The Southern Ute Tribe's Growth Fund has also grown since its inception in 1999; it currently manages more than $3 billion in assets and is the county's largest employer ("Economic Snapshot" 2017).

However, as measured by enplanements at the Durango–La Plata County Airport in July 2016, tourism a year after the GKM spill reached its lowest monthly level since January 2012 ("Tourism Data" 2017). In terms of tourism measured via the D&SNGR, ridership for July 2016 was also at its lowest level since at least 2002 ("Tourism Data" 2017). Thus a mixed portrait emerges regarding tourism-generated revenue in La Plata County following the 2015 GKM spill. A thorough examination of economic impacts following the spill is offered in chapter 6.

CONCLUSION

This chapter began with a brief overview of Colorado's history and a broad introduction to the Animas River in the state's southwestern corner. This portion of the Animas River's basin was then divided in two—the Upper and Lower watersheds; their boundaries were defined and the physical attributes specific to each were presented. For the Upper watershed and later the Lower, introductory discussions centered on the human histories, contemporary demographics, politics, local governments, and economics unique to San Juan and La Plata Counties and their administrative seats of Silverton and Durango, respectively.

For both watersheds, Native American occupation involved various bands of Utes, followed by Spanish explorers and ultimately the arrival of prospectors and hardrock miners in the early 1860s. Silverton became the hub of mining in the Upper watershed, where the Eureka Mining District was established. A number of high-yielding mines were patented (i.e., the Sunnyside Mine group) in the Cement Creek drainage immediately north of Silverton. In the Lower watershed, Durango became the supply hub and, later, milling location for the mined mineral ores. Naturally linked by the Animas River, the two towns were mechanically connected with completion of the D&SNGR in 1882.

Cycles of boom and bust typical of mining regions followed, as did development of a unique political culture rooted in Populism, which championed both the rights of working-class miners and the free coinage of silver as legal tender. With its support of the early labor movement, government reform, and economic regulation, the Populist movement in mining regions such as San Juan and La Plata Counties was a natural ally for the emergent Progressive Party and, ultimately, the Democratic Party. This explains why voters in communities such as Silverton and Durango have long since had an affinity with Democratic and, to a lesser degree, reform-minded third-party candidates.

While this common pattern of partisan support is to be expected given the silver mining legacies of the two counties, it is somewhat unusual when considering their modern demographic profiles. The two counties differ greatly in population sizes, levels of income and education, and economic sectors. Perhaps these differences are manifest more through local government and officials in response to local political issues.

Broad discussion of this issue is found in chapter 10 of this volume, particularly in terms of public and political views on Superfund designation in the Upper watershed as a means to remediate the ecological damage from AMD in the entire Animas and San Juan River basins. Many political observers have long since maintained that local elections and politics matter more, especially in terms of direct impacts on our everyday lives. For decades, disagreement among different populations on how best to deal with AMD from abandoned mine sites has been pronounced. In the context of the 2015 GKM spill and the resultant multi-state plume of pollution, this was particularly evident. Moreover, the type of response and the preferred course of action following the disaster remain contested among different communities affected by the spill.

NOTES

1. The Colorado Mineral Belt traverses the state diagonally, from Boulder to Durango.

2. Dr. John Kessell, e-mail message to author, November 11, 2017. It is common for residents of southwestern Colorado to refer to the river as el Rio de las Animas Perdidas—the River of Lost Souls. This is most likely a misnomer. Rivera's personal journal, dated July 4, 1765, contains no mention of the "lost souls" part of the river's name. Somewhat less common is reference to the Animas River as El Rio De Las Animas Perdidas en Purgatório—the River of Souls Lost in Purgatory. Rivera's journal does not reference this name either. A possible source of these two misnomers may be related to Colorado's other Animas River, in present-day Las Animas County in the state's southwest corner. According to historian David Lavender (mentioned in Kessell's email), a group of Spanish explorers stumbled on the remains of an earlier Spanish exploration group in the 1600s on the banks of a river. This prompted them to name the river el Rio de las Animas en Purgatório. Today, the shortened name of this river is the Purgatoire. Today in La Plata County, there is a Purgatory Creek, which is a tributary to Cascade Creek, located just west of

the Purgatory Resort. Perhaps an early explorer of the region, who had previously been to or been in contact with someone from southeastern Colorado, brought this name to southwestern Colorado. After extensive research, no definitive source of how the references to "lost souls" or "Purgatory" made their way to the Animas River watershed was found.

3. The prospecting party of Charles Baker first constructed the bridge above a narrow gorge over the Animas River, roughly 15 miles north of present-day Durango, in the early 1860s. Its placement in popular culture was solidified by the well-known scene from the 1968 film *Butch Cassidy and the Sundance Kid*, when the two stars jumped from the bridge into the Animas below.

4. Ty Churchill, e-mail message to author, November 8, 2017.

5. Christian S. Robbins, email message to author, December 7, 2017.

6. In 2015, San Juan County employed a staff of twenty-four (including elected officials).

REFERENCES

"About Us." 2016. ARSG (Animas River Stakeholders Group). http://animasriver stakeholdersgroup.org/blog/index.php/about-us/.

"American National Election Studies—Cumulative File (1948–2004)." 2017. American National Election Studies. http://www.electionstudies.org/studypages/down load/datacenter_all_NoData.php.

Blair, Rob, ed. 1996. *The Western San Juan Mountains: Their Geology, Ecology, and Human History*. Niwot: University Press of Colorado.

"Board of County Commissioners." 2017. La Plata County, Colorado. http://co .laplata.co.us/government/board_of_county_commissioners.

Brace, Stephen, and Vincent Virga. 2009. *Colorado: Mapping the Centennial State through History*. Guilford, CT: Globe Pequot.

Brasch, Sam. 2015. "The Gold King Mine: From an 1887 Claim, Private Profits and Social Costs." Colorado Public Radio. http://www.cpr.org/news/story/gold -king-mine-1887-claim-private-profits-and-social-costs.

"Budgets/Audits." 2017. Town of Silverton, Colorado. https://www.colorado.gov /pacific/townofsilverton/budgetsaudits.

Burbank, Wilbur, and Robert Luedke. 2017. "Geology and Ore Deposits of the Eureka and Adjoining Districts, San Juan Mountains, Colorado." Washington, DC: US Government Printing Office. https://pubs.usgs.gov/pp/0535/report.pdf.

Cassells, E. Steve. 1997. *The Archaeology of Colorado*. 2nd ed. Boulder: Johnson Books.

"Colorado." 2017. GEOSTAT. http://www.geostat.org/CO.

"Colorado Elections, by County." 2017. *Denver Post*, November 9. http://data.denver post.com/election/results/county/.

Colorado Encyclopedia. 2017. "San Juan County." http://coloradoencyclopedia.org/article/san-juan-county.

"Comprehensive Annual Financial Report for the Year Ending December 31, 2016." 2017. City of Durango Finance Department. http://www.durangogov.org/ArchiveCenter/ViewFile/Item/293.

"Comprehensive Economic Development Strategy." 2016. San Juan County, Colorado. http://www.scan.org/uploads/7_-_San_Juan_County_Update_2016.pdf.

Congressional Research Service. 2020. *Federal Land Ownership: Overview and Data*. Washington, DC: Congressional Research Service. https://fas.org/sgp/crs/misc/R42346.pdf.

Coughlin, Christine W., Merrick L. Hoben, Dirk W. Manskopf, and Shannon W. Quesada. 1999. "A Systematic Assessment of Collaborative Resource Management Partnerships." Master's thesis, University of Michigan, Ann Arbor.

"County Employment and Wages in Colorado—Fourth Quarter 2016." 2017. US Department of Labor. https://www.bls.gov/regions/mountain-plains/news-release/pdf/countyemploymentandwages_colorado.pdf.

Dean Runyan Associates. 2016. "Colorado Travel Impacts: 1996–2016." Colorado Come to Life. http://www.deanrunyan.com/doc_library/COImp.pdf.

Dell' Amore, Christine. 2009. "Ancient Camels Butchered in Colorado, Stone Tools Show?" *National Geographic*. https://news.nationalgeographic.com/news/2009/02/090227-ice-age-tools.html.

"Economic Snapshot 2017." 2017. Region 9 Economic Development District of SW Colorado. http://www.scan.org/uploads/Region_9_Final_complete_document.pdf.

Evans, Mark E. 2010. "Early Exploration of the San Juan Region." Silverton Branch of the Durango & Rio Grande. http://www.narrowgauge.org/ncmap/excursion3_silvrton.html.

Frodeman, Robert. 2003. *Geo-Logic: Breaking Ground between Philosophy and the Earth Sciences*. Albany: State University of New York Press.

"Geology and Ore Deposits of the Eureka and Adjoining Districts, San Juan Mountains, Colorado." 2017. US Department of the Interior. https://pubs.usgs.gov/pp/0535/report.pdf.

"Government." 2017a. The Official Website of San Juan County, Colorado. http://www.sanjuancountycolorado.us/government.html.

"Government." 2017b. Town of Silverton, Colorado. https://www.colorado.gov/pacific/townofsilverton.

Gustafson, Daniel. 2003. "Hydrological Unit Project." Montana State University, Bozeman. http://www.esg.montana.edu/gl/huc/.

Hamm, Kevin. 2016. "Map: Colorado Voter Registration by County." *Denver Post*, October 4. https://www.denverpost.com/2016/10/04/map-colorado-voter-registration-by-county/.

Johnston, Richard, Michael G. Hagen, and Kathleen Hall Jamieson. 2004. *The 2000 Presidential Election and the Foundations of Party Politics*. New York: Cambridge University Press.

Kazin, Michael. 2006. *A Godly Hero: The Life of William Jennings Bryan*. New York: Alfred A. Knopf.

Leip, David. 2017. "Dave Leip's Atlas of US Presidential Elections." http://www.uselectionatlas.org.

"Mining Companies, Operators, and Owners in Colorado." 2017. *The Diggings* (Chattanooga, TN). https://thediggings.com/entities-in-usa-co.

National Park Service. 2008. "Early Mining and Transportation in Southwestern Colorado: 1860–1881." https://www.nps.gov/parkhistory/online_books/blm/co/10/chap6.htm#27.

"National Register of Historic Places: Colorado–La Plata County." 2017. National Register of Historic Places. http://www.nationalregisterofhistoricplaces.com/co/La+Plata/state.html.

Ninnemann, John L., and Duane A. Smith. 2006. *San Juan Bonanza: Colorado's Mining Legacy*. Albuquerque: University of New Mexico Press.

Noel, Thomas. 2015. *Colorado: A Historical Atlas*. Norman: University of Oklahoma Press.

"Oil and Gas." 2017. Fort Lewis College, School of Business Administration. https://www.fortlewis.edu/ober/oilandgas.aspx.

Painter, Nell Irvin. 2008. *Standing at Armageddon: A Grassroots History of the Progressive Era*. New York: W. W. Norton.

Pendley, William Perry. 2017. "Gold King: EPA's Two-Year Rolling Disaster and a Path Forward to Fix It." Mountain States Legal Foundation. https://www.mountainstateslegal.org/docs/default-source/default-document-library/gold-king-and-superfund-final.pdf?sfvrsn=bbed0ac2_0.

"The Permanent Settlement of Southwestern Colorado." 2008. National Park Service. https://www.nps.gov/parkhistory/online_books/blm/co/10/chap7.htm.

Postel, Charles. 2007. *The Populist Vision*. New York: Oxford University Press.

"Region 9 Economic Quarterly." 2017. Fort Lewis College, School of Business Administration. https://www.fortlewis.edu/ober/Region9CountyData/TouristData.aspx.

"Region 9 Employment Data Update." 2016. Region 9 Economic Development District of SW Colorado. http://www.scan.org/uploads/Region_9_Employment_Trends_Updates_2016_-_2.pdf.

"Regional Data." 2017. Region 9 Economic Development District of SW Colorado. http://www.scan.org/index.php?regional-data=yes.

Rockler, Harmen. 2013. "Why American Voters Decide to Vote for Third Parties in Presidential Elections." Syracuse University Honors Program Capstone Projects 52, Syracuse, NY. https://surface.syr.edu/honors_capstone/52.

Seyfarth, Jill, and Ruth Lambert. 2010. "Pioneers, Prospectors, and Trout: A Historic Context for La Plata County, Colorado." State Historical Fund Project Number 2008-02-012, Deliverable No. 7. http://laplata.hosted.civiclive.com/UserFiles/Servers/Server_1323669/File/La%20Plata%20County's%20Community%20Development%20Services%20Department%20Migration/Planning/Historic%20Preservation/PioneersProspectors_Trout_Jan2010.pdf.

Smith, Duane A. 1996. "The Miners: 'They Builded Better than They Knew.'" In *The Western San Juan Mountains: Their Geology, Ecology, and Human History*, edited by Rob Blair, 7–8. Boulder: University Press of Colorado.

Smith, Duane A. 2007. *Durango Diary II*. Durango, CO: Herald Press.

"Tourism as a Base Industry: Update 2015." 2015. Region 9 Economic Development District of SW Colorado. http://www.scan.org/uploads/Tourism_Industry.pdf.

"Tourism Data." 2017. Fort Lewis College, School of Business Administration. https://www.fortlewis.edu/ober/Region9CountyData/TouristData.aspx.

"Town of Silverton: Budget and Financial Trends." 2016. Colorado Department of Local Affairs. https://www.colorado.gov/pacific/townofsilverton/budgets audits.

"2017 Budget: City of Durango, Colorado." 2017. City of Durango. http://www.durangogov.org/DocumentCenter/View/3394.

US Census Bureau. 2017a. "Durango City, Colorado." https://www.census.gov/searchresults.html?page=1&stateGeo=none&searchtype=web&cssp=Typeahead&q=Durango+city%2C+CO&%3Acq_csrf_token=undefined.

US Census Bureau. 2017b. "La Plata County." https://www.census.gov/searchresults.html?q=La+Plata+County%2C+CO&page=1&stateGeo=none&searchtype=web&cssp=Typeahead&%3Acq_csrf_token=undefined.

US Census Bureau. 2017c. "Silverton Town, Colorado." https://www.census.gov/searchresults.html?page=1&stateGeo=none&searchtype=web&cssp=Typeahead&q=Silverton+town%2C+CO&%3Acq_csrf_token=undefined.

US Department of Commerce. 2017. "Community Facts." https://factfinder.census.gov/faces/nav/jsf/pages/community_facts.xhtml?src=bkmk#.

US Department of Labor. 2017. "County Employment and Wages in Colorado—Fourth Quarter 2016." https://www.bls.gov/regions/mountain-plains/news-release/pdf/countyemploymentandwages_colorado.pdf.

US Geological Survey. 2017. "Current Conditions for Colorado: Streamflow." https://waterdata.usgs.gov/co/nwis/current/?type=flow&group_key=huc_cd.

Velde, François R., and Warren E. Weber. 2017. "A Model of Bimetallism: Working Paper 588." Federal Reserve Bank Research Department. https://www.minneapolisfed.org/research/wp/wp588.pdf.

2

The Gold King Mine Release

Impacts on Water Quality and Aquatic Life

SCOTT W. ROBERTS

The Gold King Mine (GKM) release, depicted in photos of the discolored Animas River, captured local, regional, national, and even international attention. Among many people's first questions were, What is in the water, and Did it kill all life in the river? This chapter addresses these two core ecological questions by evaluating the water quality of the GKM plume and relaying the findings of studies conducted to assess the impacts of the release on fish and aquatic insects. A discussion of these topics must be couched within an understanding of the geographic context of the Animas River watershed where the GKM release occurred, as well as the conditions of the Animas prior to the release.

GEOGRAPHIC AND HISTORICAL CONTEXT: WATER QUALITY AND AQUATIC LIFE IN THE ANIMAS RIVER

The Animas River drains a highly mineralized area of the San Juan Mountains, resulting in water quality that is greatly influenced by naturally occurring

DOI: 10.5876/9781646421756.c002

metals and minerals. It is believed that early explorers found some reaches of Animas River tributaries to have such high naturally occurring metal concentrations that they were largely devoid of aquatic life. The names of waterways near Silverton, Colorado—for example, Cement Creek and Mineral Creek—point to this historical legacy. Exacerbating natural mineralization, water quality in the watershed has also been impacted by the legacy of hardrock mining that occurred in the 1800s and 1900s (Besser et al. 2008). Although there is very little active mining in the San Juan Mountains today, abandoned mine tunnels honeycomb the mountains at the headwaters of the Animas watershed, and waste rock and mine tailings remain scattered throughout the region. These legacy remains of historical mining contribute metals to aquatic systems through continuous discharge from mine adits and percolation through waste rock and tailings during runoff events and annual snowmelt. While research continues to further differentiate the sources of metals (i.e., natural mineralization versus legacy mining), high levels of metals in stream water and sediments have the potential to alter aquatic biota.

It is well documented that a number of reaches in the Animas River watershed near Silverton had limited aquatic life due to high levels of metals well before the 2015 GKM release (Anderson 2008; Besser and Brumbaugh 2008; EPA 2015; Roberts 2016a). Cement Creek, a tributary to the Animas River along which the GKM is located, was and remains fishless and largely devoid of aquatic life except for sparse populations of metal-tolerant aquatic insects such as *EukiefeRiella* midges. The 30 mile reach of the Animas River from the confluence with Cement Creek in Silverton to Bakers Bridge is inhabitated by low abundances of non-native trout and a depauperate aquatic insect community. As the Animas emerges from the San Juan Mountains into the Animas River Valley through the City of Durango and on its way to the confluence with the San Juan River in New Mexico, metal concentrations gradually decrease and aquatic life is exposed to additional stressors including excessive nutrients, sedimentation, low flows, and elevated water temperatures.

Non-native fish species are also prevelant throughout the Animas River watershed, but they become more abundant in the Lower Animas, often displacing native fish. In the reach of the Animas near Durango, fish populations are sustained through stocking of non-native rainbow and brown trout by Colorado Parks and Wildlife (CPW). This annual stocking program is a managmeent intervention to support the lucrative recreational fishery, since natural reproduction is limited by poor water quality and habitat conditions (White 2015). It is unlikely that cold-water fish species such as trout would persist in the Lower Animas at the abundance they do today without stocking.

AQUATIC LIFE AND MINE-IMPACTED WATERSHEDS

The acidic drainage that accompanies mine-impacted watersheds can affect aquatic life, namely fish and benthic macroinvertebrates (BMIs), in myriad ways. BMIs are small organisms that live in rocks and sediment at the bottom of rivers and lakes; they are a critical component of aquatic food chains. Their diversity and wide range of habitat requirements, food sources, life spans, and tolerances to pollution make them excellent indicators of water quality and stream health. Aquatic invertebrates are particularly susceptible to metal contamination, as they are readily exposed to metals in the water column as well as in river sediments. Furthermore, metals can directly and indirectly impact aquatic life. Iron, for example, can cling to the gills of fish and BMIs, thus inhibiting respiration. Aquatic organisms can inadvertently consume metal-contaminated sediment and suspended particles or uptake metals dissolved in solution (Luoma 1983). Precipitation of metals on river-bottom substrate can indirectly affect aquatic life by smothering potential habitat, filling interstitial spaces, and preventing the growth of aquatic vegetation, which some BMIs rely on as a food source (Vuori 1995).

In the upper portion of the Animas River watershed, impacts on aquatic life from the legacy of historical mining are well-known and have been extensively researched (Anderson 2008, 2010; Courtney and Clements 2002; Roberts 2017b; Smith 1976). In a study conducted prior to the GKM release, the Environmental Protection Agency (EPA 2015) evaluated the potential risks to the environment from metal contamination of water and sediments in the Upper Animas watershed. Their research concluded that due to elevated metal concentrations, fish have largely been eliminated in the Animas River from Silverton to the confluence with Elk Creek and that benthic communities were impaired in the Animas River from Silverton to Baker's Bridge. In another study, researchers gathered benthic insects from Elk Creek, a tributary to the Animas River with good water quality, and placed them in plastic chambers in the Animas River just upstream from Elk Creek for a period of thirty days. Researchers found that a significant number of BMIs placed in the Animas River died. In particular, Heptageniidae and Ephemerellidae mayfly and Taeniopterygidae stonefly abundances were substantially reduced after exposure to Animas River water (Courtney and Clements 2002).

Although the distribution of aquatic life in the Upper Animas River is largely shaped by concentrations of metals in water and sediment, changes in metal concentrations over time have directly translated to fluctuations in the diversity and abundance of aquatic life. A number of mine remediation activities occurred in the Upper Animas watershed during the 1990s and early 2000s, including the temporary operation of a mine wastewater treatment plant, stabilization of mine waste rock, and the installation of bulkheads. A treatment plant was operated on Cement Creek upstream of Silverton from 1978 to 2004 to remove metals from water draining the American Tunnel and, at times, all of Cement Creek (ARSG 2017). Following the closure of the plant, there was an increase in the volume of water draining the nearby Red and Bonita Mine and a potentially related decline in water quality in Cement Creek (ARSG 2017). Concurrently, diversity of the benthic community in Animas River sites downstream of Silverton decreased (Anderson 2010; Roberts 2017a). Further downstream in the Animas River canyon, a family of mayflies, Ephemerellidae, which is known to be especially sensitive to elevated metals, became absent and the density of brook trout and rainbow trout declined (Roberts 2017a; White 2010, 2015). Aquatic life in the Animas River watershed has also responded positively to reductions in metals associated with mine remediation efforts. In Mineral Creek, a tributary to the Animas River, remediation over the past two decades has resulted in large reductions of copper, cadmium, and zinc. In 2016, biologists with the US Geological Survey unexpectedly found brook trout in a reach of Mineral Creek previously known to be devoid of fish (EPA 2019; Romeo 2016).

THE GKM RELEASE AND WATER QUALITY

The GKM release caused a visible plume of metal-laden water to enter Cement Creek; over a period of nine days, it flowed down the Animas and San Juan Rivers to its terminus in Lake Powell Reservoir (EPA 2017). The plume carried water and materials that had been backed up behind the collapsed GKM portal, along with waste rock and sediment mobilized downstream of the mine. The US Environmental Protection Agency (EPA 2017) estimated that 1 percent of the metals in the GKM plume were directly discharged from the GKM tunnels, while 99 percent were from waste rock material that was located below the mine portal and was mobilized by the erosive force of the event.

By the color alone, it was apparent that the plume had a different chemical makeup than typical Animas River water. The plume, an opaque mustard color, contrasted with the typically clear, green, or brown color of the Animas River. Newspapers across the country carried headlines remarking on the bizarre color (see chapter 3 for detailed discussion of the impact of this dramatic discoloration).

Along with this change in color, the plume was generally associated with a drop in pH, a rise in conductivity, elevated turbidity, and a large increase in metal concentrations. The distinct yellow color was caused primarily by oxidation of iron and aluminum in the plume that formed colloidal solids (EPA 2017). The character of the plume changed as it moved downstream. Solubility of metals is dependent on several factors, most notably pH, which can dictate whether a metal is in a dissolved or particulate form. Dissolved metals are more readily assimilated by aquatic life and thus are generally considered more harmful than particulate metals. The EPA estimated that most of the metal load carried by the plume into the Animas River was in a colloidal or particulate state rather than in a dissolved form. As the plume traveled downriver, it became diluted by other waters, raising the pH and causing geochemical reactions that led to the deposition of metals on the streambed (EPA 2017). Modeling of the plume by the EPA indicates that much of the metal load carried by the plume was deposited along the Animas River channel prior to reaching the San Juan River.

Water quality samples collected by the Mountain Studies Institute (MSI) from the Animas River at Rotary Park in Durango show that when the plume passed, there was a more than 500 percent increase in levels of aluminum, arsenic, cobalt, copper, iron, lead, manganese, silver, vanadium, and zinc (table 2.1) (Roberts 2016b). In some instances, concentrations of metals measured in the plume were high enough to be of concern for the health of aquatic life. Various states and tribes have established acute water quality criteria to protect aquatic life from brief, short-term exposure to contaminants during events like the GKM release. As the plume moved downstream,

TABLE 2.1. Change in metal concentrations as the GKM release passed through Durango, Colorado

Metal (µ/L)	*Pre-Release (August 6, 2015, 8 p.m.)*	*Peak of Release (August 7, /2015, 12:30 a.m.)*	*Direction of Change*	*Percent Change*	*Pre-Release (August 6, 2015, 8 p.m.)*	*Peak of Release (August 7, 2015, 12:30 a.m.)*	*Direction of Change*	*Percent Change*
	Total Recoverable				*Dissolved*			
Aluminum	122	12,300	Increase	9,982	59.4	20*	Decrease	-66
Antimony	2.5*	10.9	Increase	336	0.5*	0.5*	None	–
Arsenic	2.5*	87.5	Increase	3,400	0.6	0.5*	Decrease	-17
Barium	43.4	208	Increase	379	50.6	22.1	Decrease	-56
Beryllium	0.15*	0.20	Increase	33	0.15*	0.15*	None	–
Cadmium	0.5*	2.85	Increase	470	0.139	0.699	Increase	403
Calcium	53,100	66,600	Increase	25	51,200	62,700	Increase	22
Chromium	5*	7.85	Increase	57	2.12	1.27	Decrease	-40
Cobalt	0.5*	5.12	Increase	924	0.26	1.66	Increase	536
Copper	2.5	395	Increase	15,512	2.55	4.32	Increase	69
Iron	152	121,000	Increase	79,505	100*	100*	None	–
Lead	1.5	2,620	Increase	175,739	3.3	0.119	Decrease	-96
Magnesium	7,210	11,100	Increase	54	7,020	7,930	Increase	13
Manganese	90.1	1,330	Increase	1,376	75.3	676	Increase	798
Mercury	0.05*	0.26	Increase	410	No data	No data	–	–
Molybdenum	5*	25.8	Increase	416	1*	1*	–	–
Nickel	2.5*	2.5*	None	-	0.5*	0.5*	–	–
Potassium	1,920	5,410	Increase	182	1,830	2,020	Increase	10
Selenium	5*	6.9	Increase	38	1*	1*	None	–
Silver	2.5*	16.3	Increase	552	0.5*	0.5*	–	–
Sodium	10,600	10,900	Increase	2.8	10,200	10,400	Increase	2.0
Thallium	2.5*	11.6	Increase	364	0.5*	0.5*	–	–
Vanadium	10*	60.8	Increase	508	2*	2*	–	–
Zinc	58	980	Increase	1,590	10.0	84.8	Increase	748

* Denotes samples with metal concentrations below the method detection limit (MDL). In these cases, the MDL is listed (Roberts 2016b).

aluminum was the only metal that exceeded acute aquatic life standards at multiple locations along the Animas and San Juan Rivers. Other metals in the plume either did not exceed acute water quality standards or, in the case of cadmium, copper, lead, manganese, and zinc, only exceeded standards in the stretch of the Animas River upstream of Durango (EPA 2017).

Fortunately, the duration of the plume and associated increase in metal concentrations was relatively brief. Metal concentrations in the Animas and San Juan Rivers had returned to pre-plume levels within one to two weeks following the incident (EPA 2017; Roberts 2016b). However, concerns remain that increased flows during monsoonal storm events and spring runoff could remobilize sediments deposited by the GKM release and cause additional spikes in metal concentrations. This concern was evaluated by the MSI by collecting water quality samples from the Durango stretch of the Animas River during multiple storm events in the late summer and fall of 2015 and during spring runoff in 2016. The MSI found that metal concentrations did increase during these events but were at levels consistent with historical data for the Durango stretch of the Animas River. Conversely, in the New Mexico stretch of the San Juan River, the EPA reported that aluminum concentrations detected during storm events and spring runoff after the GKM release persisted at levels above historical observations. The mobilization of Gold King–related metals during the 2016 spring runoff was predicted by EPA modeling, which concluded that almost all metals that had been deposited along the Animas River channel during the GKM release were transported downstream to Lake Powell by the end of the 2016 spring runoff period (EPA 2017).

THE GKM RELEASE AND AQUATIC LIFE

The GKM plume contained metals that, when they occur at a high enough concentration and for a long enough duration, can be toxic to aquatic life. Aluminum concentrations in the plume surpassed levels known to be acutely harmful to aquatic life (EPA 2017; Roberts 2016b). Two separate studies were conducted to evaluate potential impacts on aquatic life from the GKM release; CPW reported on fish communities and MSI reported on benthic communities. Researchers examined potential impacts on aquatic life from the GKM release by evaluating organism survival immediately after the release, examining tissue metal concentrations, and comparing community composition observed during the fall of 2014—prior to the GKM release—to those observed during the fall of 2015 and 2016, following the release.

On August 6, prior to the arrival of the GKM plume in Durango, CPW placed live rainbow trout (*Oncorhynchus mykiss*) in cages in the Animas River at three locations near Durango. After five days, CPW removed the cages and found that 207 of the 208 trout had survived exposure to the toxic plume (White 2015). CPW also studied potential impacts on fish populations by comparing results from electrofishing surveys conducted prior to and after the GKM release at several sites on the Animas River from Silverton to Durango. At a site in the Animas River canyon known to be only sparsely populated by trout, CPW reported a slight increase in the density of brook trout (*Salvelinus fontinalis*) following the GKM release. Further downstream, in the Durango stretch of the Animas River, CPW found an increase in rainbow and brown trout (*Salmo trutta*) biomass and density after the GKM event. In addition to trout, CPW also examined potential impacts on populations of mottled sculpin (*Cottus bairdii*), a fish native to the Animas River known to be particularly sensitive to contaminants, including dissolved zinc (Besser and Brumbaugh 2008; Brinkman and Woodling 2005). A month after the GKM release, CPW found robust mottled sculpin populations in the Durango stretch of the Animas River; densities were higher than the historical average, and there was clear evidence of recent reproduction (White 2015). Metal concentrations in fish tissue collected from the Animas River in the fall of 2015 were found to be at comparable levels to fish in other Colorado rivers (CDPHE 2015). These results suggest that the GKM release did not have a measurable impact on Animas River fish populations (White 2015).

Similarly, the MSI evaluated the survival of BMIs by comparing samples collected from the Animas River immediately prior to the GKM release to samples collected twenty-four hours and one week after the release. Within the immediate seven days following the GKM release, the MSI found no evidence that aquatic insect communities in the Durango stretch of the Animas River were negatively impacted. The MSI also found that the diversity and abundance of aquatic insects did not decrease immediately following the GKM event (Roberts 2016a).

In addition to assessing immediate survival of BMIs in the Durango stretch of the Animas River, the MSI was able to evaluate potential impacts on benthic communities at a broader spatial scale using data collected in 2014 from thirteen sites in the Animas watershed from Silverton to Durango. By repeating benthic surveys in 2015 following the GKM release and using the same survey methodology, sites, and seasonal timing as data collected in 2014, the MSI could directly compare pre- and post-GKM release BMI populations at sites scattered over a 50 mile stretch of the Animas River. Eight of the thirteen sites were on the Animas River downstream of GKM and thus were exposed to the plume. The remaining five sites were either tributaries to the Animas River or were located upstream of where the plume entered the Animas River and thus were used as control sites to differentiate any release-related impacts from any region-wide phenomena unrelated to the GKM release. Upon examining metrics of community composition (such as diversity, abundance, and sensitive species), the MSI found no statistically significant difference in Animas River aquatic insect communities between 2014 and 2015, indicating that the GKM release did not consistently alter aquatic insect community health or structure at surveyed sites in the Animas River (figure 2.1). However, it is important to again highlight the historical context of water quality in the Animas River because the 2014 pre-spill data in this study already did not contain BMIs that might have been most affected by an event like the GKM release. There are a few families of BMIs that are known to be particularly sensitive to metal contamination. These include two families of mayflies (Heptageniidae and Ephemerellidae) and one family of stoneflies (Taeniopterygidae) (Courtney and Clements 2002; Iwasaki et al. 2009; Roberts 2017b). Historical data indicate that these metal-sensitive families were either absent or only occurred at low densities in many stretches of the Animas River, demonstrating that benthic communities were likely already impacted by metal contamination prior to the GKM release. The MSI found no evidence that small populations of metal-sensitive families were impacted at any site, whether through extirpation or a large reduction in density, following the GKM release (Roberts 2016a).

These survival data only paint a partial picture of impacts on aquatic life from contamination. It is also critical to examine the potential for non-lethal effects that could result in diminished fitness or unsuccessful reproduction. One method of assessing potential non-lethal impacts is to determine whether contaminants are accumulating in insect body tissue, which can result in cascading accrual up through higher trophic levels. At the same

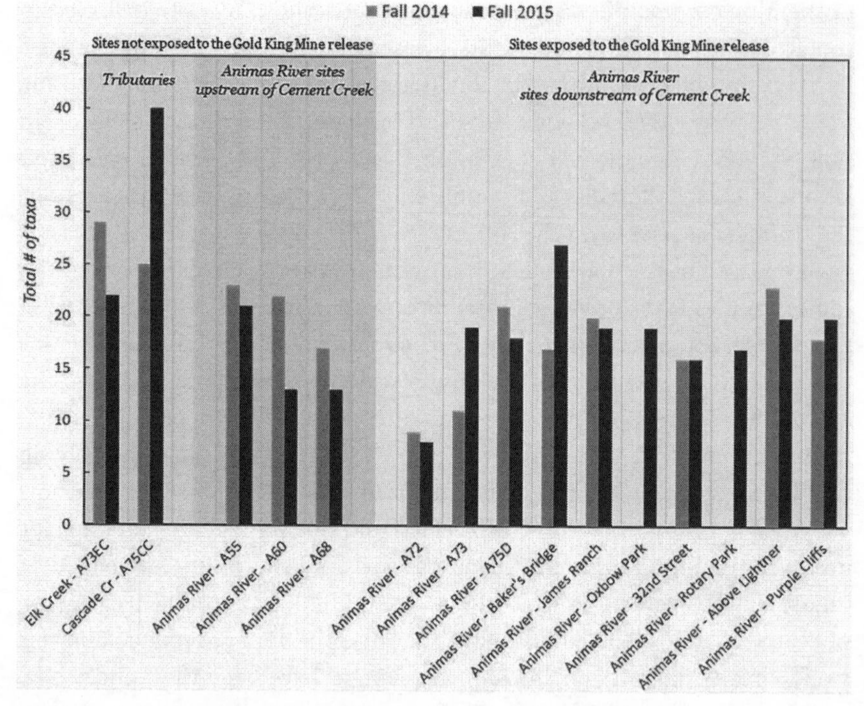

FIGURE 2.1. Benthic taxa richness before (2014) and after (2015) GKM release (Roberts 2016a)

thirteen locations where the MSI collected BMI community samples, the MSI also gathered BMI tissue samples to be analyzed for metal concentrations. Comparing 2014 to 2015, the GKM release is associated with a pronounced increase of iron and aluminum in BMI tissue concentrations at the Animas River site located closest to where the GKM plume entered the Animas River and a statistically significant increase in copper tissue concentrations at all sites exposed to the GKM plume (figure 2.2; table 2.2). The higher and more persistent increase in copper tissue concentrations relative to other metals could be attributable to the high proportion of copper that has historically occurred in water draining the GKM. Prior to the GKM release, the US Geological Survey (USGS) estimated that approximately 40 percent of the copper load from the Silverton area originated from the GKM (Runkel et al. 2019). Data collected one week after the GKM release from Cement Creek indicate that dissolved copper increased more than any other metal when

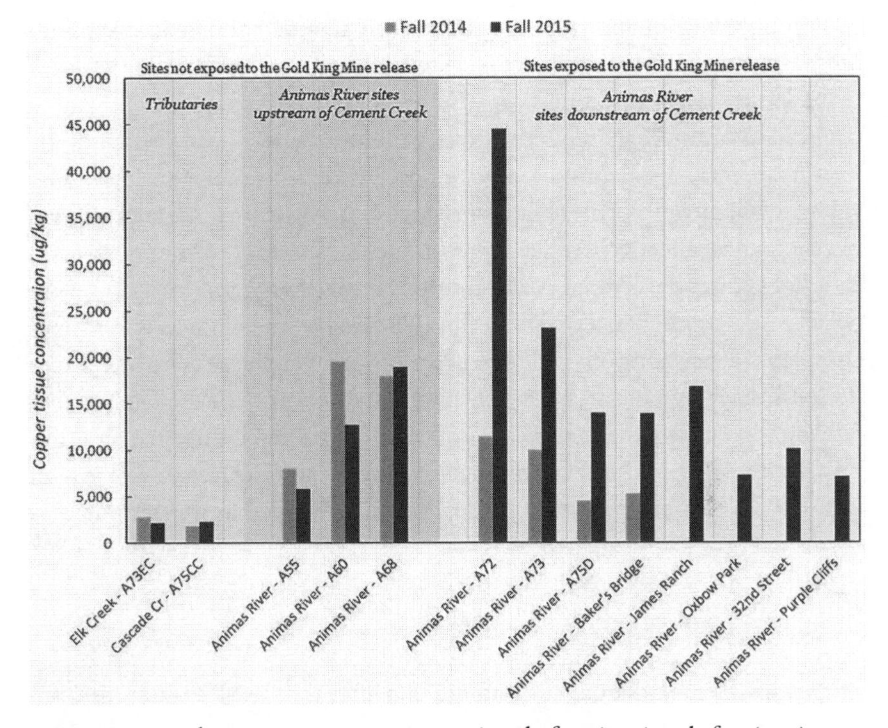

FIGURE 2.2. Benthic copper tissue concentrations before (2014) and after (2015) GKM release (Roberts 2016a)

TABLE 2.2. Mean BMI tissue metal concentration for sites exposed to the GKM plume (Roberts 2016a)

Metal	*Fall 2014 Mean*	*Fall 2015 Mean*	*Difference*	*P-value (based on paired t-test)*
Aluminum	176.1	246.0	69.9	0.16
Cadmium	299.5	271.3	-28.3	0.64
Copper	7,820.0	23,875.0	16,055.0	0.03
Iron	574.5	805.5	231.0	0.15
Lead	1,435.0	1,230.0	-205.0	0.70
Manganese	44.3	22.9	-21.4	0.99
Zinc	67.9	77.9	10.0	0.28

compared to observations from 2009–2014 (Runkel et al. 2019). Similarly, copper in stream sediment in the Animas River below Silverton increased more than any other metal following the 2015 spill (Runkel et al. 2019).

Although there did not appear to be any lethal impacts on benthic communities from the GKM release, the elevated levels of iron and copper in insect tissue led researchers to question whether sub-lethal impacts occurred that could have affected subsequent insect generations. It is known that some benthic insects are tolerant of elevated metal concentrations during the larval stage but are intolerant at later life stages, such as during metamorphosis or emergence, effectively reducing reproduction success (Wesner et al. 2014). To address this concern, the MSI surveyed benthic communities again in the fall of 2016—more than a year after the GKM event, when insects would have completed at least one life cycle since exposure to the plume. The MSI found no statistically significant difference in any BMI metric between fall 2014 and fall 2016, indicating that the release did not consistently alter BMI community health or structure at surveyed sites in the Animas River one year after the spill—which suggests that it did not cause sub-lethal impacts such as reduced reproduction success.

CONCLUSION

After assessments of aquatic life before and after the GKM release, researchers concluded that despite brief exposure to elevated metal concentrations, the release does not appear to have caused extirpation of fish or aquatic insects. It is perhaps surprising that an event of the magnitude of the GKM release did not have a more dramatic impact on aquatic life (see chapter 3 for further discussion).

These findings should be interpreted within the historical context of chronic metal contamination in the Animas watershed. Fish and benthic communities in the Animas River have long been stressed by exposure to mine-related impacts and other stressors (Anderson 2008; EPA 2015; Smith 1976). Prior to the GKM release, trout reproduction in the Durango stretch of the Animas River was poor, and aquatic insects most sensitive to metal contamination were diminished or absent altogether (Roberts 2016a; White 2015). Because of these degraded conditions, it is possible that aquatic life in the Animas River may have a diminished capacity to exhibit a response to stressors such as those present in the GKM plume (Rehn et al. 2011; Roberts 2016a). In other words, the aquatic life present in the Animas River on the day of the GKM release constituted the species already able to survive periodic exposure to contaminants, and the GKM release did not appear to *further* impact the already distressed aquatic life of the Animas River.

REFERENCES

Anderson, Chester. 2008. "Effects of Mining on Benthic Macroinvertebrate Communities and Monitoring Strategy." Washington, DC: US Geological Survey Professional Paper.

Anderson, Chester. 2010. *2010 Benthic Macroinvertebrate Data Analysis*. Durango, CO: Animas River Stakeholders Group.

ARSG (Animas River Stakeholders Group). 2017. "Animas River Stakeholders Group: Cement Creek Historical Timeline." http://animasriverstakeholdersgroup.org/blog/index.php/2015/10/23/gold-king-timeline/.

Besser, John M., and William G. Brumbaugh. 2008. "Toxicity of Metals in Water and Sediment to Aquatic Biota." Washington, DC: US Geological Survey Professional Paper.

Besser, John M., Susan E. Finger, and Stanley E. Church. 2008. "Impacts of Historical Mining on Aquatic Ecosystems: An Ecological Risk Assessment." Washington, DC: US Geological Survey Professional Paper.

Brinkman, Stephen, and John Woodling. 2005. "Zinc Toxicity to the Mottled Sculpin (*Cottus bairdi*) in High-Hardness Water." *Environmental Toxicology and Chemistry* 24 (6): 1515–1517.

CDPHE (Colorado Department of Public Health and the Environment), ed. 2015. *Colorado Department of Public Health and the Environment, Analysis and Recommendation: Fish Consumption—Animas River*. Denver: Colorado Department of Public Health and the Environment.

Courtney, Lisa A., and William H. Clements. 2002. "Assessing the Influence of Water and Substratum Quality on Benthic Macroinvertebrate Communities in a Metal-Polluted Stream: An Experimental Approach." *Freshwater Biology* 47 (9):1766–1778. doi: 10.1046/j.1365-2427.2002.00896.x.

EPA (Environmental Protection Agency). 2015. *Baseline Ecological Risk Assessment, Upper Animas Mining District, San Juan County, Colorado*. Washington, DC: EPA.

EPA (Environmental Protection Agency). 2017. *Analysis of the Transport and Fate of Metals Released from the Gold King Mine in the Animas and San Juan Rivers*. Washington, DC: EPA.

EPA (Environmental Protection Agency). 2019. *Aquatic Baseline Ecological Risk Assessment, Bonita Peak Mining District Superfund Site, San Juan County, Colorado*. Washington, DC: EPA.

Iwasaki, Yuichi, Takashi Kagaya, Ken-Ichi Miyamoto, and Hiroyuki Matsuda. 2009. "Effects of Heavy Metals on Riverine Benthic Macroinvertebrate Assemblages with Reference to Potential Food Availability for Drift-Feeding Fishes." *Environmental Toxicology and Chemistry* 28 (2): 354–363. doi: 10.1897/08–200.1.

Luoma, Samuel N. 1983. "Bioavailability of Trace Metals to Aquatic Organisms: A Review." *Science of the Total Environment* 28 (1–3): 1–22. doi: 10.1016/S0048-9697(83)80004-7.

Rehn, Andrew, Peter Ode, and James Harrington. 2011. *The Effects of Wildfire on Benthic Macroinvertebrates in Southern California Streams*. San Diego, CA: San Diego Regional Water Quality Control Board.

Roberts, Scott. 2016a. *Mountain Studies Institute Animas River 2015 Benthic Macroinvertebrate (BMI) Report, Gold King Mine Release Monitoring*. Silverton, CO: Mountain Studies Institute.

Roberts, Scott. 2016b. *Mountain Studies Institute Animas River Water Quality at Rotary Park, Durango, Colorado: Gold King Mine Release Monitoring*. Silverton, CO: Mountain Studies Institute.

Roberts, Scott. 2017a. *Mountain Studies Institute Animas River 2017 Benthic Macroinvertebrate Assessment*. Silverton, CO: Mountain Studies Institute.

Roberts, Scott. 2017b. *Mountain Studies Institute Bonita Peak Mining District 2016 Benthic Macroinvertebrate Assessment*. Silverton, CO: Mountain Studies Institute.

Romeo, Jonathan. 2016. "Trout Discovered in Creek Long Devoid of Fish." *Durango Herald*. https://durangoherald.com/articles/93990.

Runkel, Robert, Daniel Cain, and Scott Roberts. 2019. "Metal Concentrations in Bed Sediments and Benthic Macroinvertebrates, Before and After the 2015 Gold King Mine Release (Colorado, USA)." Vienna, Austria: European Geosciences Union.

Smith, Norwin F. 1976. *Aquatic Inventory, Animas–La Plata Project: Final Report/by Norwin F. Smith*. [Denver, CO?]: Colorado Division of Wildlife. Monograph.

Vuori, Kari-Matti. 1995. "Direct and Indirect Effects of Iron on River Ecosystems." *Annales Zoologici Fennici* 32 (3): 317–329.

Wesner, Jeff S., Johanna M. Kraus, Travis S. Schmidt, David M. Walters, and William H. Clements. 2014. "Metamorphosis Enhances the Effects of Metal Exposure on the Mayfly, Centroptilum triangulifer." *Environmental Science and Technology* 48 (17): 10415–10422. doi: 10.1021/es501914y.

White, Jim. 2010. *Colorado Parks and Wildlife 2010 Animas River Report*. Denver: Colorado Parks and Wildlife.

White, Jim. 2015. *Colorado Parks and Wildlife 2015 Fishery Inventory Report for the San Juan and Upper Dolores River Basins*. Denver: Colorado Parks and Wildlife.

3

A Potent Focusing Event

The Gold King Mine Spill and Rapid Policy Development

BRAD T. CLARK

This chapter posits the 2015 Gold King Mine (GKM) blowout and resultant toxic spill as a focusing event and examines the significant policy development and change that rapidly ensued in the aftermath. Focusing events often involve unexpected crises; due to their stunning and often dramatic nature, intense media coverage and widespread emotional reactions have been linked to subsequent political action, followed by significant policy development and policy change in the aftermath (e.g., Birkland 1997, 2007).

The GKM event occurred on August 5, 2015; less than four months later, the San Juan County commissioners and Silverton Town Board of Trustees reversed twenty-five+ years of opposition and voted unanimously to direct city staff to pursue a Superfund listing. Following an additional nine+ month period, the US Environmental Protection Agency (EPA) officially listed the Bonita Peak Mining District (BPMD) on the National Priorities List (NPL) under the Superfund Program on September 9, 2016. It took just over a year for this significant policy development to materialize following multiple decades of policy stasis and inaction, maintained through opposition by area residents, as well as local, state, and national politicians.

DOI: 10.5876/9781646421756.c003

This chapter also assesses the likelihood of additional policy development and change to other, long-established regulatory policies in the areas of hardrock mining and water quality—specifically the 1872 General Mining Law, the 1972 Clean Water Act, and the 1980 Superfund itself. Efforts to amend these landmark US laws to more effectively address the problems of acid mine drainage (AMD) predate the GKM spill; all prior efforts at reform have been unsuccessful. The potential for future ecological and social harms from abandoned hardrock mines remains a real and growing concern in thousands of watersheds across the West. The chapter concludes by assessing the politics and policy dilemmas that may impact future efforts to effectively deal with the impacts of historical mining activities in the American West.

FOCUSING EVENTS

According to literature from the field of political science and related study of public policy, focusing events are sudden, dramatic, and relatively rare occurrences that simultaneously attract the attention of policymakers and the general public (e.g., Birkland 1997, 1998, 2007; Kingdon 2003). These events often involve crises, including natural or human-made disasters, terrorist attacks, or commercial airline crashes. Their dramatic, crisis-oriented images provide those seeking policy change with opportunities to convey their perceptions of policy failures to broader audiences and ultimately bolster the attempts of pro-change groups or coalitions to pursue change in the status quo for a given policy area. Policy change, however, can take quite different forms and magnitudes. Following in the path of Thomas A. Birkland (2007), the main consideration in this chapter is actual policy change after disaster, not just agenda setting. The policy option of Superfund designation in the Upper Animas watershed had been on the agenda since the mid-1990s, yet persistent NIMBYism combined with local preference for nonfederal, stakeholder-led remediation efforts effectively prohibited policy development in the direction of comprehensive federally led cleanup. The following are a variety of examples of substantive policy change following a notable crisis-oriented focusing event.

After the 1989 Exxon *Valdez* oil spill in Alaska's Prince William Sound, the US Congress passed the Oil Pollution Act of 1990, which called for the phasing out of single-hull tankers and mandated double-hull tankers in all US waters by 2010. This represents the type of policy change defined by the *adoption* of new legislation. Prior to this, in 1969, the then largest oil spill in US waters occurred off the California coast in the Santa Barbara Channel. In the fifty+ years since the dramatic event, no new leases for offshore drilling have been issued by the California State Lands Commission within its jurisdiction—out to the 3 nautical mile limit. This represents the type of policy change defined by the *termination* of existing programs or policy. In a third context, the unexpected 1981 discovery of arsenic in the drinking water of Bonner, Montana, from the centuries-long accumulation of heavy metals behind Milltown Dam on the Clark Fork River prompted the 1983 listing of the largest Superfund complex in the western United States. This represents the type of policy change defined by the substantive *development* or *advancement* of existing legislation. The EPA's 2016 listing of the BPMD outside of Silverton, Colorado, following the 2015 GKM spill is most similar to this latter context of policy change. In the spill's aftermath, the BPMD's official listing on Superfund's NPL materialized in a notably rapid manner—just over twelve months following the spill.

The GKM Blowout as Focusing Event and Media Sensation

Regarding both visual and psychological impacts, the August 2015 blowout at GKM and the ensuing toxic plume that grossly discolored a multi-state river were extreme. Locally, coverage of the spill occurred almost daily, with the *Durango Herald* publishing approximately 275 news reports and updates between August 5, 2015, and August 27, 2016. The event was also covered by major print outlets including the *Denver Post, Time*, *Newsweek*, the *Wall Street Journal*, the *New York Times*, the *Washington Post*, and the *Los Angeles Times*.[1] Of note, the GKM event was listed by *Newsweek* magazine among the top five "Nastiest Cases of Toxic Discharge in 2015." The piece began with this sensational statement: "A river in Colorado turned a bright mac-and-cheese orange with mining waste" (Schlanger 2016).

The GKM story also received primetime national and international television coverage from all major outlets, and several internet news organizations contributed to its broad coverage. The spill even impacted popular culture, with an appearance on Conan O'Brien's TBS show *Conan*, in the form of a fake ad for the ultimate extreme sport: "The Colorado Natural Disaster Ultimate River Rapids Extreme Kayaking Adrenaladventure Tours" (Butler 2015).

In the days and weeks following the GKM event, journalists from multiple outlets took their cue from the term *discolored*, describing the plume of AMD in a range of near apocalyptic terms. For example, two days following the spill, the *Washington Post* used the phrase "a million gallons of filthy yellow mustard" in an article's title to describe the incident (Kendall-Ball 2015). Another outlet used the phrase "a puke-colored plume of mine runoff" (Pendergast 2015).

This type of intense, sensationalized media coverage has drawn the attention of many who specialize in the study of media and communications. For example, consider the following interpretation of the GKM event and resulting media coverage from Pulitzer Prize–winning author Michael Kodas:

> Of course, most simply, the color made for such a dramatic image. Anyone, anywhere in the world, even if they didn't speak English, could look at that spectacularly scenic river and see that something wasn't right. Visuals drive so much news coverage anymore. Even more importantly, the fact that the EPA has become such a political football in the last couple of years has allowed several presidential candidates not generally renowned for their environmental leanings to attack the EPA as environmental protectors. Particularly for the international media, which is fascinated by the reality TV show of American presidential politics, they found all the outlandish statements about the Animas River in this political sideshow as colorful as the incident itself. Also, there's a worldwide mythology of the American West, from the different landscapes to the miners, cowboys and rugged individuals who settled the West, which really fascinates countries that are centuries or millennia past that stage in their own history. (quoted in Pendergast 2015)[2]

According to Birkland (1997, 2007), a leading scholar of focusing events and policy change, the initial reaction to an event is most readily detectable through the news media—especially the twenty-four-hour news cycle, which makes sudden, unexpected, and calamitous events particularly attractive for ongoing coverage. Birkland further asserts that environmental disasters are much more likely to yield highly photogenic stories when compared to events that require a compilation of data and indicators, and perhaps changes in magnitude, to convey problem severity. The GKM event produced an incredibly photogenic story, with a handful of powerful images that quickly went viral around the globe (figures 3.1, 3.2).

FIGURE 3.1. Hazmat suit–clad kayaker on the Animas in Durango. *Courtesy*, Steve Fassbinder.

FIGURE 3.2. Post-spill Animas River, Durango, Colorado. *Courtesy*, EcoFlight/Bruce Gordon.

The media's role is especially important for this chapter, since increased levels of coverage have been shown to correspond with greater levels of public and political attention to problems (Birkland 1997, 2007; Smith 1992). Further, the social construction of stories, metaphors, and symbols often serves to advance (or limit) the movement of issues on both public and political agendas (Stone 1997). Social construction refers to the general process by which societies collectively define and explain the nature and cause of problems (Stone 1997). Since journalists use symbols (i.e., images) and stories (i.e., narratives) to frame and explain complex issues and events, the public—as news consumers—responds to particular images and stories as it interprets the nature and causes of problems.

Regardless of nationality or language spoken, everyday citizens intuitively recognized that a dayglow-orange–colored river is not normal but rather indicative of something very unusual and unnatural. The focality of the mine spill confirmed that the discolored river had a central or definitive point of origin—the lowermost Level-7 adit (or portal) at GKM just north of Silverton, Colorado. While some focusing events derive their power and influence from objective attributes of a particular event (e.g., gallons spilled, chemicals released, lives lost, or people affected), media portrayal and social construction may enhance and focus this power by reducing focusing events to simple, unnerving, or graphic images that are symbolic of larger, more complex problems (Birkland 1997). The focal power of a surreally discolored Animas River ultimately proved greater than the numerical indicators associated with water quality, acidity, and fish mortality following the spill. In addition, the symbolic propagation of the orange river was greatly facilitated by the intense international media coverage that converged on the seemingly pristine river flowing through a scenic mountain valley in a secluded corner of the southwestern United States.

Another way to gauge relative interest in major events comes from data available via Google, Inc., which allows a unique perspective on what people are interested in or concerned about at particular points in time. Specifically, *Google Trends* shows how often a particular search term is entered relative to the total search volume at certain times and across regions. For the year of the GKM event, the search term "2015 Gold King Mine wastewater spill" yielded the highest possible value during the week following the disaster (August 9–15). This value of 100 refers to the peak popularity of the term. For a few weeks, values for the same search term steadily declined until reaching the single digits for the remainder of the year, following the week of September 20–26.

From a regional perspective, peak popularity for the same search term was evidenced most in the State of New Mexico (value 100), followed by Colorado (value 67), Arizona (value 35), Utah (value 28), and the District of Columbia (value 16). While this last value suggests that the search term was less than 20 percent of the peak popularity for New Mexico, it nevertheless shows that individual searches in the District of Columbia (DC) were higher than in all other US states, save for the four most directly impacted by the GKM spill. This relative spike in DC-area searches was likely the result of interest from policymakers, congressional aides, agency officials, and other political actors in the nation's capital.

Finally, from a relative perspective, the same GKM search term yielded the highest possible value for the week of August 9–15 when compared to the only other natural resource– and disaster-related news event covered on August 5, 2015.[3] Compared to the peak popularity of the GKM spill (value 100), news on an oil spill off the California coast (value 5) prompted a mere 5 percent of the searches relative to GKM's peak popularity. People were clearly interested in obtaining online information about the GKM disaster.

Visual versus Environmental Impacts (images versus indicators)

According to mainstream scholars in areas of agenda setting and policy change, systemic or numerical indicators are critical for garnering and expanding public and political attention to certain problems (e.g., Baumgartner and Jones 1993, 2002; Birkland 1997, 2007; Kingdon 2003). Such indicators are used to assess both the magnitude of and the change in problem severity. Yet regarding events that transpired in the months following the GKM spill, indicators of environmental impact and degradation in the form of water quality and aquatic habitat destruction were not the driving force behind the significant policy change. Rather, perceived problem severity was largely a function of visual images of the discolored river and the ensuing narratives perpetuated through intense media coverage.

The aforementioned account demonstrates how the GKM spill was a truly *visual* focusing event that garnered widespread attention. It began, literally, with a forceful blowout. The ensuing saga surrounding the visibly tarnished river and associated fears about public and environmental health persisted for months. For many area residents, such concerns remain to this day.

Nonetheless, three years after the GKM event, the general consensus provided by water quality data and biological indicators remains the same—ecological damages to the Animas and San Juan Rivers were largely temporary and not as severe as originally feared. Regarding water quality, tests conducted in Durango during the plume's peak found five exceedances of water quality standards for metal concentrations. Specifically, aluminum levels exceeded the Aquatic Life Acute Standard as set by the Colorado Department of Public Health and the Environment (CDPHE); and levels of arsenic, iron, lead, and thallium exceeded the EPA's Recreational Screening Level. Yet ever since the initial plume passed, no state or federal water quality exceedances have been detected ("Gold King Mine Spill" 2016).

In anticipation of the plume's arrival in Durango during the first days following the spill, the Colorado Department of Parks and Wildlife (CDPW) placed 108 fingerling trout—measuring 1½ to 2 inches in length—in pens along the river in Durango in order to evaluate impacts. After enduring the full effect of the toxic plume, the fish remained in place for the following week and were monitored daily by biologists. With the exception of one that reportedly died immediately from unrelated causes, all the fish survived, prompting a spokesperson from the CDPW to state: "Our biologists were really surprised. That plume of acid runoff looked pretty bad. Survival, honestly, wasn't expected" (Benjamin 2015).

The fish were subsequently collected and taken to Denver, where they were dissected and tested for metal accumulations in tissue and internal organs. Test results were inconclusive, suggesting no significant evidence of heavy metals loading. For the most part, this has been attributed to the fact that the Animas River has been impacted by AMD for more than a century (and by acid rock drainage [ARD] for millennia); while this long-term exposure has been detrimental to fish populations, it is likely that resident populations have developed tolerance to local conditions and increased resiliency (Benjamin 2015). Despite the jolt of metals released by the spill, it was a fast-moving event, which inhibited deposition of a significant blanket of metals on the riverbed (Benjamin 2015).

Only seven days after the plume passed Durango, the CDPHE authorized officials in Durango to reopen intake structures for public water systems; three days later, the New Mexico Environment Department lifted the ban on San Juan County's drinking water system supplied by the Animas and San Juan Rivers.[4] Less than two weeks after the spill, the La Plata County sheriff reopened the Animas to recreation ("Drinking Water Impact Questions" 2015).

Although the relatively fast-flowing Animas River worked to flush a large portion of the heavy metals released from GKM in a timely manner, the EPA's longer-term concern is the effect on the entire watershed from metals deposited in sediments and their release during high-water events and from recreational use over time. The EPA has indicated that these sediments may pose some risk, especially to aquatic life and fish ("Frequent Questions" 2016). In addition, both the CDPHE and the EPA recommend that water from wells located in the Animas floodplain be regularly tested for possible contamination, especially along the slow-flowing segments just north of Durango.

The fact that the GKM spill was followed with such rapid and substantive policy change in the form of NPL Superfund designation—even though numerical indicators of ecological conditions suggested minimal impact—is remarkable. Moreover, the paucity of significant environmental harm, discussed below in more detail, was surprising to many biologists. Indeed, longtime scholars of the policy process have made a direct connection between problem severity and the rate and extent of related policy change (e.g., Baumgartner and Jones 1993, 2002; Birkland 1997, 2007; Kingdon 1984). As discussed throughout this chapter, the GKM case is unique; Superfund listing was achieved not because of calamitous ecological damage and an abundance of numerical indicators but rather as a result of the unprecedented visual impact and emotional responses caused by the spill and broadcast widely through national and international media outlets.

Despite the less than expected extent of environmental impacts following the spill, the legacy of the visual impact propagated by the intense media coverage exceeded most expectations. It has taken longer for the associated emotional impacts to diminish; indelible images of a grotesquely discolored river flowing through a picturesque landscape remain frightening and still conjure feelings of anger and disbelief.

At face value, the release of 3 million gallons of contaminated AMD into an alpine watershed sounds extreme—and it is when caused by a single event.

Yet each year, abandoned mines in the Silverton area collectively discharge an estimated 330+ million gallons of AMD into the Animas River watershed—an amount more than 100 times the volume of the 2015 spill (Pratt 2017). As incredible and unbelievable as it may sound, the Animas River experiences an AMD release on par with that from the GKM event every three-odd days. The only difference is that it is less concentrated and thus does not produce a dramatic discoloration of water. Visitors who ride the Durango & Silverton Narrow Gauge Railway to Silverton each summer may notice discolored trails of iron hydroxide deposits in the Animas just below Silverton, known locally and in mining circles as "yellow boy," or in Cement Creek as it flows through the town's park. These peculiarities are isolated, however, and pale in comparison to an entirely discolored multi-state river.

GKM AS CATALYST FOR POLITICAL LEARNING AND POLICY DEVELOPMENT—NPL LISTING

Superfund

At the broadest level, Superfund is the federal government's program to clean up the nation's abandoned and uncontrolled hazardous waste sites. It is operated under legislative authority from the Comprehensive Environmental Response, Compensation, and Liability Act of 1980 (CERCLA) and the 1986 Superfund Amendments and Reauthorization Act (SARA). Together, these statutes provide the basis for EPA implementation of solid waste emergency recovery and long-term removal and remedial activities ("Superfund" 2016).

In large part, congressional passage of Superfund in 1980 represents the decades-long culmination of public and political pressures from a series of controversies (i.e., focusing events) involving public health risks from abandoned hazardous waste sites—all of which were among the inaugural Superfund sites listed in September 1983 (Colten and Skinner 1996; Levine 1982). The best-known of these occurred at the Love Canal neighborhood in Niagara Falls, New York, where municipal refuse and, later, 20 tons of toxic waste from chemical manufacturing had been dumped for decades prior to the neighborhood's construction—in an area where more than 1,000 homes and apartments and two public schools were ultimately built. Significantly elevated rates of miscarriages, birth defects, and cancers among area residents led to the listing of Love Canal as the Hooker (S) site; after two decades and $400 billion in remediation actions, it was formally de-listed in 2004.

Two other hazardous waste sites have similarly been characterized as catalysts for Superfund's passage (Colten and Skinner 1996; Levine 1982). One was located near the community of Times Beach, Missouri, where various companies had produced the chemical herbicide and defoliant Agent Orange, which was used extensively in Vietnam. The chemical hexachlorophene—an antibacterial agent—was also produced onsite, and large quantities of heavily contaminated used motor oil were also stored and used for dust suppression on local roadways. The cumulative result was the most severe dioxin contamination in US history, necessitating the eventual abandonment of 800 residential properties, as well as the listing of the 180 acre Syntex Superfund site. It was removed from the NPL in 2001. The other site—unofficially known as the Valley of the Drums—was located near Louisville, Kentucky. It had been used to store 100,000 drums of toxic liquid waste since the early 1960s. Leakage due to inadequate storage practices polluted nearby aquifers and surface waters with a host of heavy metals and known carcinogens called polychlorinated biphenyls (PCBs). In response, the Distler Farm Superfund site was designated in 1983; cleanup activities ended in 1990, yet the site remains actively listed on the NPL.

Originally, Superfund was financed by federal taxes collected from industrial petrochemical companies. The tax-based trust fund reached a peak of $3.8 billion in the early 1990s, but the financing mechanism expired in 1995 and was never reauthorized by the US Congress; its balance was effectively gone by 2003. Today, Superfund projects rely on congressional appropriations and, when possible, monies collected from so-called potentially responsible parties (PRPs). Superfund is thus partially based on the *polluter pays principle*, whereby the federal government attempts to identify all PRPs for a given site; based on the degree of their financial viability, PRPs may be financially responsible for all or some portion of overall cleanup costs (Churchill 2017). When the EPA is unable to identify financially viable PRPs, the federal government finances cleanup and restoration activities at the given site for ten years, after which the state in which the site is located assumes all expense and permitting responsibilities (Churchill 2017). In the context of the GKM spill, PRPs may include former and current property and mine owners and mine operators. As of mid-2018, no PRPs had officially been identified. It is likely that at some point, the owner of a neighboring mine, the Sunnyside Gold Corporation, and its parent company, the Kinross Gold Corporation—the world's fifth largest gold producer—will be listed as significant PRPs.

The National Priorities List and Hazardous Ranking System

The NPL is the primary guide for the EPA in determining which waste sites warrant further investigation. The NPL lists the sites deemed most pressing among the known or threatened releases of hazardous substances, pollutants, and contaminants throughout the United States and its territories ("Superfund: National Priorities List" 2016). As required by CERCLA, the EPA updates the NPL annually; this provides policymakers with a list of high-priority sites and demonstrates the size and nature of the nation's ongoing cleanup challenges.

Upon proposing the addition of an NPL site, the EPA publishes a public notice in the Federal Register and makes announcements through local media outlets to notify affected communities; this affords concerned community members the opportunity to comment on a proposed listing. If, at the end of the formal comment period, a site still qualifies for cleanup, it is formally listed on the NPL. Once listed, the EPA publishes a notice in the Federal Register and responds formally to comments it received. In addition, the EPA may issue a fact sheet or flyer to notify the communities most impacted by a site's listing ("Superfund: National Priorities List" 2016).

The EPA uses a Hazardous Ranking System (HRS) to assess the relative threat to human health and the environment from the actual or potential release(s) of hazardous substances. Data from HRS reports describe the primary sources of contamination (i.e., "pathways") and explain the basis for a site's ranking and rationale for NPL designation. The HRS is therefore the primary screening tool for determining whether a site qualifies for inclusion on the NPL; to be eligible, a site must receive a minimum HRS score of 28.5. Scores range from 0 to 100; these scores do not represent a specific level of risk but rather a cutoff point that "serves as a screening-level indicator of the highest priority releases or threatened releases" ("Superfund: National Priorities List" 2016).[5]

The area that ultimately became the BPMD received an HRS Site Score of 50.0, based on scoring in one (surface water) of the possible contamination

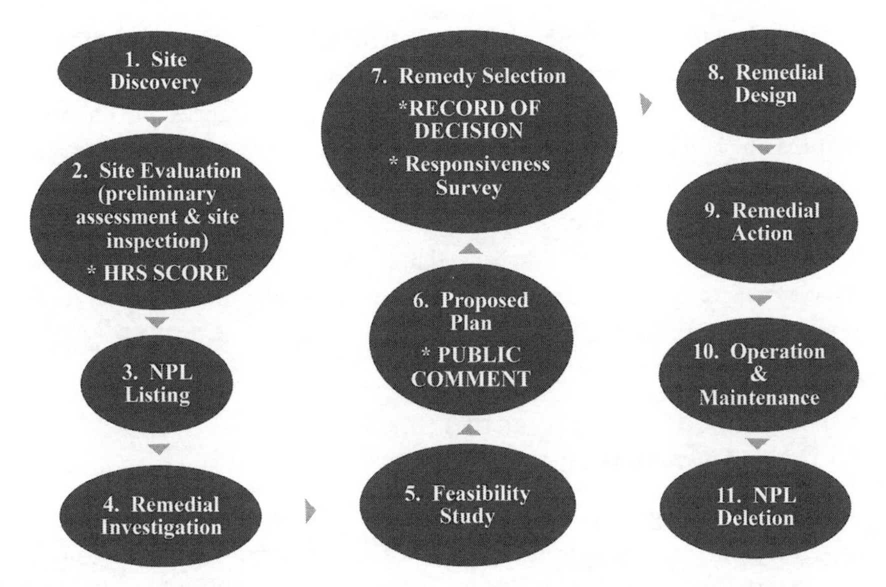

FIGURE 3.3. The Superfund process. *Courtesy*, "Superfund Proposed Plan," US Environmental Protection Agency, accessed July 31, 2017, https://semspub.epa.gov/work/04/10664405.pdf

pathways evaluated in the HRS process (air, groundwater, soil). Of the twenty current NPL sites in Colorado, only five received higher scores, including the infamous Rocky Flats site, where components for nuclear weapons were fabricated and assembled from 1952 to 1975, and the Rocky Mountain Arsenal site, where the US Army manufactured chemical weapons and munitions from 1942 to 1952 and, later, where the Shell Chemical Company produced pesticides and herbicides until 1982 ("National Priorities List [NPL] Sites by State" 2017).

The entirety of the Superfund process involves eleven sequential, uniform phases. It is thus considered a linear process (Churchill 2017). After a site is identified, the (second) Site Evaluation Process occurs, which broadly characterizes the degree of risks to public health and the environment, after which the actual HRS score is calculated. Official NPL listing comes next, which is followed by the (fourth) Remedial Investigation phase, where the nature and extent of contamination at the site is characterized. This is followed by the (fifth) Feasibility Study phase, which is essentially a cost-benefit analysis of the treatment options identified for the NPL site. The EPA then proposes its Interim Remedial Actions Plan (IRAP) for inaugural restoration work at the site in the (sixth) Proposed Plan stage. This is followed by the (seventh) Remedy Selection phase, where the EPA announces any changes that may have resulted in the preceding phase; any changes to the IRAP are based primarily on consideration of public comments and ongoing feasibility studies. These official plans to remedy threats to public health and the environment at the NPL site are formalized in the Records of (remedy) Decision. As of May 2019, restoration activities at the Bonita Peak Mining District were transitioning into the (eighth) Remedial Design phase.

SUPERFUND COMES TO SILVERTON . . . OR, RATHER, TO THE "BONITA PEAK MINING DISTRICT"

The EPA first began formal consideration of Superfund status for the Silverton, Colorado, area in the early 1990s; ever since, many of the town's approximately 650 full-time residents have expressed vocal opposition. Notwithstanding, the EPA increased its data collection in 2008 to assess whether GKM and other upper Cement Creek mines qualify for Superfund listing. Although no official determination was made, the EPA announced in 2010 that the Upper Animas would likely satisfy listing criteria for Superfund.

In addition to a profound distrust of the federal government—the EPA in particular—and related fears of the overall loss of local control, opposition has centered on a combination of NIMBY-inspired apprehensions regarding possible stigmatization of the tourist-dependent area as dangerous or unsafe to visit. Local business and community leaders tout the concomitant reductions in sales and tax revenues as reason to avoid federal listing. In 2014, two service sectors of San Juan County's economy (i.e., arts, entertainment, recreation and accommodation, food services) accounted for 67 percent of jobs and 57 percent of all income ("Comprehensive Economic Development Strategy" 2015.). In addition, existing mine owners fear being held liable as PRPs for reimbursing the tens of millions of dollars in costs the EPA would incur with comprehensive remediation; moreover, many long-time residents would genuinely like to see industrial-scale mining return to the Silverton area (Langolis 2015).

Opposition to Superfund designation has also centered on fears associated with decreased property values. While the apprehension over property values is valid, existing research has shown that property values may actually *increase* following comprehensive cleanup and subsequent de-listing of a Superfund site. Specifically, a peer-reviewed study found that residential property values within 3 miles of a Superfund site rose 18.6–24.5 percent post-remediation and subsequent removal from the NPL ("Superfund Remedial Annual Accomplishments" 2017). Support for such a conclusion can be gleaned from the Smuggler Mountain Superfund site in Aspen, Colorado, de-listed in 1999. Today, multi-million-dollar homes occupy the area once contaminated with heavy metals from hardrock silver and lead mining.

Potential Superfund designation rose again on the Silverton agenda in April 2014, when EPA officials repeated their proposal to list parts of Silverton and the surrounding area, without which there would be no money for long-term cleanup efforts. The EPA also declared that Superfund designation would allow further study into the extent of AMD in the Upper Animas Basin. The GKM blowout was roughly a year away. When it ultimately happened, on August 5, 2015, everything changed.

After decades of steadfast resistance, local politicians and community members faced a new reality. The release of 3 million gallons of contaminated water from GKM and the dramatic color change of the Animas brought intense national and international attention to the Silverton community and to the broader issue of AMD across the western United States. After considering the alternatives, the Town of Silverton's Board of Trustees and the San Juan County commissioners voted unanimously on November 23, 2015, to instruct town officials to pursue Superfund listing by initiating discussions with the EPA and the State of Colorado.

As the next step in the process, Silverton's seven-member Board of Trustees and the three county commissioners were scheduled to vote on whether to approve a draft letter in support of Superfund designation to Colorado governor John Hickenlooper (D) during the week of January 25, 2016. The decision was postponed, however, due to a number of unresolved concerns among local residents: clarification of the site's actual boundaries, concerns over reimbursement costs potentially incurred by the Town of Silverton, and (re)assurances that any negative impacts associated with cleanup would be mitigated (Paul 2016).

To address these concerns, all parties agreed to a revised site definition for NPL listing. Originally, the EPA proposed including the entire Upper Animas

River drainage, including all tributaries. This would have covered more than 50 percent of San Juan County, as well as the entire Town of Silverton and thousands of acres not directly associated with water quality issues in the watershed. The EPA agreed to an amended site definition that included only mining claims reasonably believed to be directly associated with contamination in the Animas, thereby protecting the town itself and hundreds of mining claims that posed no known environmental hazards.

Naming of the new site was also a concern—mainly for area residents because the EPA originally proposed names that might unnecessarily draw adverse attention to the Town of Silverton and San Juan County. A name was thus agreed upon that did not connote anything negative about the town or the county; the official name of the cleanup area emerged as the Bonita Peak Mining District—named after the peak into which the GKM was cut. Lastly, the EPA agreed to ensure local involvement in all phases of the cleanup process following Superfund designation and to furthermore adopt a process through which any properties initially listed within the site could be removed if additional studies found they did not materially contribute to contamination of the Animas and its tributaries ("Town of Silverton News" 2016).

On February 29, 2016, Colorado governor John Hickenlooper officially notified the EPA Region 8 administrator that he supported the proposed listing of the BPMD on Superfund's NPL, thus making it eligible for federal cleanup and assistance. Along with the GKM, the governor's letter advocated the targeted cleanup of forty-five other mines and two settling pond areas in the Upper Animas watershed—for a total of forty-eight specific sites. This executive support from the state level represents the final step in the process for a community to formalize its request for NPL listing. Two months later, the EPA accepted Hickenlooper's request and officially proposed the BPMD site on the Federal Register. What had previously been hotly contested and vehemently opposed for decades was now a reality, approved unanimously by town and county officials, Colorado's governor, and the EPA nearly eight months to the day after the initial GKM blowout. According to San Juan County commissioner and former opponent of Superfund designation Scott Fetchenhier, "I think history has been made. This is one of the most important decisions ever made by [the] county commissioners or town council" (quoted in Hood 2016).

A similarly upbeat statement on Superfund listing came from Mike Esper, editor of the *Silverton Standard and the Miner*:

> There were two big concerns about Superfund that this community had . . . One: It would kind of foreclose on the future of returning to mining. And the other one, the big one: the bad publicity. We are totally reliant on tourism at this point . . . But, the Aug. 5 blowout . . . kind of blew that argument out of the water. That game is over. We had the bad publicity by not having Superfund and by not addressing the problem that's only going to make the publicity worse. (quoted in Hood 2016)

A less positive, albeit supportive, assessment came from long-time Silverton miner and current San Juan County commissioner Ernest Kuhlman, when speaking to the ninety-plus residents who packed Silverton's Town Hall prior to the final vote: "I was not in favor of Superfund. I still don't like it. But if we don't do it, it will be done for us . . . If we don't make this move, they will, and we won't have a seat at the table" (quoted in Finley 2016). After the vote, Kuhlman stated: "It appears at this time that the [NPL] is the best way to get these mines cleaned up quickly. All of us—Silverton, San Juan, and our downstream neighbors—want something done immediately" (quoted in Finley 2016).

Political Learning

Birkland's (2007) model of event-related learning posits a focusing event followed by policy change. A critical intervening variable is said to be *political* or *policy learning* on behalf of political actors. Thus individuals are the key objects of learning in the policy process (May 1992, 338; Sabatier 1987, 664; 1991, 144). The aforementioned statements suggest (rapid) political learning on behalf of many Silverton residents and elected officials. However, in chapter 10 of this volume, the degree or extent of authenticity regarding such learning, especially among Silverton-area residents who had opposed Superfund listing for decades, is critically assessed.

It is without question, though, that such a rapid reversal in policy preference and the unanimous endorsement of Superfund listing would not have transpired in lieu of the August 5, 2015, GKM blowout and spill. Perhaps even more so, the resultant toxic plume and shockingly discolored water

FIGURE 3.4. Toxic plume approaches Durango, Colorado. *Courtesy*, EcoFlight/ Bruce Gordon

that flowed through the heart of Durango—Silverton's larger and more affluent, yet equally tourist-dependent, neighbor to the south—likely had a more significant role in prompting the unusually rapid policy development (figure 3.4). In addition, the direct impact to irrigation and drinking water on Navajo lands likely contributed to the event's negative image; the Navajo declared in their state of emergency that at least 16,000 of their people, 30,000 acres of crops, and thousands of livestock were forced to depend on the contaminated Animas and San Juan Rivers (Tribune Wire Reports 2016).

While Birkland (2007) acknowledges that response to a crisis or a disaster requires a degree of deliberation, he recognizes that available policy tools may also be selected because of extreme pressures to act quickly in such situations. To reiterate, it took only 110 days for the San Juan County commissioners and town trustees to vote unanimously to purse Superfund listing with the EPA. Such a listing had long been an option available for county and state leaders to endorse, yet in lieu of a catastrophic focusing event like the GKM blowout, they chose to rely on other strategies (i.e., pursuit of

Good Samaritan legislation) and nonfederal actors (e.g., the Animas River Stakeholders Group, State of Colorado).

Policy Change—Making It Official

Formal recognition of the BPMD site came on April 7, 2016, when the EPA's proposed listing was placed on the Federal Register. Then, after a 68 day public comment period and the subsequent addressing of all comments in a responsiveness survey, the EPA officially added the BPMD in San Juan County, Colorado, to Superfund's NPL on September 9, 2016. After decades of unyielding opposition, it took a mere thirteen months for this significant policy development to occur. Existing literature on the policy process, specifically the implications of focusing events, suggests that this is an anomaly; prevailing literature urges caution when attempting to attribute policy change to any one cause or event (e.g., Birkland 2007). The GKM case is an exception to this rule.

ADDITIONAL AVENUES FOR POLICY DEVELOPMENT AND CHANGE

Because of the broad scope and scale of problems associated with current and future AMD, policy development at the federal level is necessary to achieve comprehensive remediation of abandoned mine sites. Indeed, since 1999, almost a dozen senators and representatives from multiple western states have supported a variety of bipartisan bills and extensions of existing law to address the issue. These lawmakers have included Max Baucus (D-MT), Michael Bennet (D-CO), Ben Nighthorse Campbell (R-CO), Cory Gardner (R-CO), Martin Heinrich (D-NM), Edward Markey (D-MA), Scott McInnis (R-CO), Ken Salazar (D-CO), Scott Tipton (R-CO), Mark Udall (D-CO), Tom Udall (D-NM), and Ron Wyden (D-OR).

Legislative efforts to formally address AMD have taken two primary directions: (1) an amendment to the 1972 Clean Water Act (CWA) to provide for a Good Samaritan liability exemption, and (2) an amendment to the 1872 General Mining Law to establish an Abandoned Mining Reclamation Fund. In this latter context, official descriptions of both US House and US Senate versions of the bill (H.R. 963 and S. 2254, respectively) prominently featured a post-GKM spill image of the Animas River. A third and perhaps more obscure context of policy change involves the extension of liability waivers to the 1980 CERCLA.

The Clean Water Act

The CWA is the primary law dealing with water quality in the United States; it charges the federal government with the restoration and maintenance of the "chemical, physical, and biological integrity of the Nation's waters." In the Animas River watershed, implementation of the CWA has been the primary driver to address AMD and mine remediation for the last twenty-five years (Churchill 2017).

The foundation for the CWA came with congressional passage of the 1948 Federal Water Pollution Control Act. In 1972, the US Congress passed the Federal Water Pollution Control Act Amendments, which brought about a significant reorganization and expansion of the original 1948 act. Together, these two have fallen under the umbrella label of the "Clean Water Act"; it provides the basic structure for regulating surface water quality standards and controlling discharges of pollutants into US waters. The EPA oversees implementation of the CWA by state-level administrative agencies, yet the federal agency retains authority to establish and enforce water quality standards and various pollution control programs, as provided in Section 309(a) of the act.

A primary objective of the act is to regulate the discharge of pollutants into US waters via a permit-based system, whereby dischargers of *point-source* pollution (e.g., those with a discrete means of conveyance such as a pipe, drain, or ditch) are required to obtain and comply with specific permitting requirements. Failure to do so is considered illegal. Under this National Pollution Discharge Elimination System (NPDES), permit holders are limited to specific types and quantities of point-source pollution discharge (i.e., effluent) and are required to maintain records and report them to the EPA. Other primary programs to regulate the discharge of effluent are found in Section 307, which requires industrial water users (e.g., indirect dischargers such as rubber or battery manufactures) to pre-treat effluent prior to its discharge into Publicly Owned Treatment Works (POTW), such as municipal treatment plants; Section 404, which requires permits from the US Army Corps of Engineers to regulate dumping of dredged or fill materials in US waterways, including wetlands; and Section 405, which requires permits to govern the disposal of sewage sludge from private treatment plants or POTWs ("National Pollution Discharge Elimination System" 2017).

Good Samaritan Liability Exemption

There is deep irony in the fact that the statutory cornerstone of water quality protection in the United States represents one of the principal obstacles to the effective treatment of AMD, yet the 1972 CWA presents an unintended consequence for parties attempting to address the problem. In this sense, the CWA created both a mandate for and an obstacle to certain types of AMD cleanup. Specifically, the CWA prohibits "the discharge of any pollutant by any person" without a permit into "navigable waters from any point source" (Limerick et al. 2005, 20). Permits are available for legal, albeit regulated, discharge under the NPDES. Since horizontal mine adits (or openings) are considered point sources of pollution, any party interested in remediation efforts must obtain an NPDES permit; these require that any treatment mechanism installed at a mine adit will yield results that meet stringent CWA quality standards. Further, the party undertaking such efforts remains legally responsible for the point source of pollution in perpetuity.

This point-source classification of water draining from mines remains the primary impediment to Good Samaritan groups like the Animas River Stakeholders Group (ARSG) from taking a more comprehensive approach to AMD. The classification of AMD as point-source pollution is also misleading; while it is entirely possible to *point* to (i.e., directly identify) a mine adit such as at the lowest level of GKM, the actual water draining from the opening is often the product of a different mine or series of mines, interconnected by human-made subterranean tunnels and natural fissures in the rock. This is especially true for the Eureka Mining District—in which GKM is located—with its combination of natural geology, annual precipitation, and extensive mining history. As a result, the estimated 526 million gallons of annual AMD from GKM and other mines in upper Cement Creek has been classified as "likely to require perpetual treatment," at an annual cost of nearly $1 million (Sumi and Gestring 2017).

This incredible liability potential has limited efforts such as ARSG's to focus on removal of waste rock and ore piles, which are considered under the CWA to be *nonpoint sources* of pollution, with no defined points of discharge. Remediation at nonpoint sources involves lower levels of liability; this allows coordinated reclamation activities by state agencies such as the Colorado Division of Reclamation, Mining, and Safety and locally driven citizen groups such as the ARSG.

When written, CWA's intent was to target factories and heavy industries that emitted pollution directly into the nation's waterways. If their owners failed to obtain an NPDES permit, they could be subject to significant financial penalties of up to $32,500 for every day they were found to be in noncompliance. Since the vast majority of hardrock mines in the West have long since been abandoned by their owners or have ownership uncertainties, cleanup efforts are often driven by third parties, including nonprofit organizations, advocacy and community groups, non-governmental organizations, private businesses, and state and local agencies. Facing such burdens, Good Samaritans have little incentive to engage in comprehensive remediation efforts aimed directly at draining abandoned mines.

Roughly five months after the GKM events, a bipartisan group of Colorado lawmakers, including US senators Cory Gardner (R-CO) and Michael Bennet (D-CO) and US representative Scott Tipton (R-CO), collaborated in formulating S. 1443—the Good Samaritan Cleanup of Orphan Mines Act, a draft of which was introduced in the Senate Committee on Environment and Public Works on January 26, 2016. If passed, the legislation would allow Good Samaritans to apply for permits from a state or Native American tribe or from the EPA to assist in the cleanup of orphan mines; their efforts would be exempt from all CWA liability. Issues that likely frustrated the act's progress include reluctance from some Democrats in Congress to reopen discussions surrounding the CWA, ambiguity over who or what could actually qualify as a Good Samaritan, and difficulties with establishing uniform benchmarks to assess the extent of cleanup (Limerick et al. 2005, 40).

As of this writing, the most current activity on Good Samaritan protections occurred on December 6, 2018, when two members of Colorado's congressional delegation introduced legislation to promote remediation of abandoned hardrock mines. Both bills were intended to "authorize the EPA to establish a pilot program under which [the agency] grants permits for Good Samaritans to remediate historic mine residue at orphan hardrock mine sites, and establish a Good Samaritan Mine Remediation Fund for land management agencies that authorize Good Samaritans to conduct remediation projects on federal land" (Congress.gov 2020).

In the US House, Representative Scott Tipton (R-CO-3) introduced H.R. 7226, the Good Samaritan Remediation of Orphaned Hardrock Mines Act of 2018. It was initially assigned to three committees: Transportation and Infrastructure, Energy and Commerce, and Natural Resources. Four days later, it was referred to the House Subcommittee on Energy and Natural Resources, where it remained, unattended to, until the end of the 115th Congress (2017–2018). In the other chamber, US senator Cory Gardner (R-CO) introduced a companion bill, S. 3727, also on December 6, with the same title. It was initially referred to the Senate Committee on Environment and Public Works, where it, too, remained untouched until the end of the 115th Congress. Both pieces of legislation had no cosponsors and thus lacked broader support from colleagues of Tipton and Gardner. The fact that both bills effectively died in committee is unsurprising, given that both lacked broad bipartisan support. Moreover, both met a predictable fate for bills introduced in the US Congress, where the vast majority die when not afforded a public reading.

The 1872 General Mining Law

The General Mining Law (GML) was passed in 1872—following the California Gold Rush and other western mining booms of the mid-nineteenth century, when mineral deposits were being explored and mine patents filed, often on federal lands. In lieu of any law governing the transfer of mineral rights from public to private ownership, prospectors were left to devise their own customs and codes (Wilkinson 1992). To codify and amend such informal agreements, the US Congress passed the GML, which provided broad discretion to the private sector over the use of resources on public lands.

The GML primarily covers extraction of precious hardrock metals (gold, silver, platinum) and minerals (aluminum, iron, copper); it does not apply to the extraction of fossil fuels such as coal, oil, and natural gas. The law required little in the way of regulation and public administration; it allowed claimants to acquire outright title to both minerals and land by obtaining a mineral patent at a per-acre cost of between $2.50 and $5.00. The GML also fixed land claims to a size of 1,500 feet long by 600 feet wide. Miners were not required to pay *any* royalty taxes to the US government on minerals and metals taken from beneath federal lands; this makes hardrock mining the sole extractive industry in the country not required to reimburse taxpayers for the wealth mined from public lands ("National Pollution Discharge Elimination System" 2017).

The GML has not been significantly updated since President Ulysses S. Grant signed it into law; mining companies continue to acquire title to public land for a few dollars per acre, pay no royalties, implement minimal environmental protections, and perform no site remediation when finished. For these reasons, the 1872 GML has been a lighting rod for controversy and the target of numerous reform attempts by elected and administrative officials.

Today, the GML's jurisdiction covers in excess of 270 million acres (or two-thirds of the public domain). Approximately $1 billion worth of hardrock minerals are mined annually from these public lands; the lost royalties are in the tens of millions of dollars (Wilkinson 1992). While the law was originally written to encourage development in the American West and to benefit small-scale prospectors, today it mostly benefits large, foreign-owned multinational corporations. According to the US Geological Survey, seven of the country's top ten gold-yielding mines are owned by foreign entities, primarily from Canada, South Africa, and Great Britain (George 2009).

Abandoned Mine Reclamation Fund?

Former president Barack Obama made reform of the 1872 GML a priority, as reflected in his annual budget proposals. Beginning with his 2012 budget request, the president included a proposal each year for a hardrock mining royalty fee as a means to facilitate cleanup of abandoned mines across the country. Each year, the proposal was eliminated by Congress.

On November 5, 2015—the three-month anniversary of the GKM event—US senator Tom Udall (D-NM) introduced S. 2254, the Hardrock Mining and Reclamation Act of 2015, in an effort to modernize the 1872 GML. The bill proposed a royalty—similar to that paid by oil, gas, and coal companies for decades under the Surface Mining Control and Reclamation Act of 1977—to help pay for abandoned mine cleanup and prevent future disasters similar to GKM. Specifically, mining companies would be charged both a royalty rate of 2–5 percent of gross income and a reclamation fee of 0.6–2 percent on annual production value from new mines to establish a Hardrock Minerals Reclamation Fund. Expectations were that it would collect nearly $100 million annually (Graham 2015). Importantly, the proposed bill only imposed the royalty and reclamation fees on new mines, thus exempting existing operations (Graham 2015). The bill had three cosponsors: Martin Heinrich (D-NM), Ron Wyden (D-OR) and Edward Markey (D-MA). After its reading before the US Senate, it was referred to the Committee on Energy and Natural Resources, where it remained until the end of the 114th Congress.

Prior to this, in February 2015, US representative Raul Grijalva (D-AZ-03) introduced similar legislation, H.R. 963—the Hardrock Mining Reform and Reclamation Act of 2015. The major difference from the Senate bill is that it included a proposed 4 percent royalty fee on *existing* mines in addition to the 8 percent royalty on new operations (Graham 2015). H.R. 963 also proposed to levy an annual reclamation fee of seven cents per ton on "displaced material" from mine operators. The bill had thirty cosponsors and a month after introduction was referred to four US House committees and two subcommittees. The last stop for H.R. 963 was the Natural Resources Committee's Subcommittee on Energy and Mineral Resources, where it remained until the end of the 114th Congress.

As with other previously unsuccessful reform attempts, these legislative efforts were stridently opposed by the mining industry, as companies continued to resist Good Samaritan bills that would effectively tax their operations to subsidize cleanup costs at others. Furthermore, the mining industry continues to advocate for legislation that would allow the practice of "re-mining" at abandoned sites to extract leftover valuable metals. Critics of this practice argue that re-mining under the guise of abandoned mine reclamation simply exchanges one environmental problem for another (Limerick et al. 2005, 25).

On September 19, 2017, Senator Michael Bennet reintroduced S. 1833, the Hardrock Mining and Reclamation Act of 2017, with the same three cosponsors. Predictably, it was referred to the Energy and Natural Resources Committee, where it died; as of this writing, no additional action has occurred. With a partisan committee makeup of twelve Republicans and eleven Democrats, the bill will likely face tough opposition, and its advancement out of committee to a chamber-wide vote is uncertain—especially given that the committee's chairwoman, Lisa Murkowski (R-AK), has a lifetime record of 19 percent when voting to support pro-environment bills. Level of support for the environment among her Republican colleagues on the committee is even lower, averaging 11 percent. The Democratic minority on the committee has a lifetime voting average of 87 percent in favor of environmental bills.[6] With Republican control of both chambers in the 115th Congress and, by design, control of all committees and subcommittees, the chances for any significant legislation to target the problem of AMD and reclamation of abandoned mines are less than favorable.

Notwithstanding, local support for Bennet's bill is much stronger. Consider these comments on the 2017 bill by elected officials ("Bennet Introduces Bill" 2017).

> La Plata County is acutely aware of the risks posed by abandoned hardrock mines and the inadequacies of the existing General Mining Law of 1872. Reform of the law is imperative to provide both resources and mechanisms to address the impacts of hardrock mining on public lands.
>
> —*Julie Westendorff, chair of the Board of Commissioners, La Plata County*

> The Gold King Mine spill, August 5, 2015, which released an orange plume of three million gallons of acid mine waste into the Animas River and through our community, provided a timely reminder that a 145-year-old law is woefully inadequate and places undo [*sic*] responsibility on the taxpayers to address the impacts of this industry on the environment, our communities and economies.
>
> —*Durango city councilor Dean R. Brookie*

> Reform of the 1872 Mining Law is long overdue, and I wholeheartedly support this bill which will give us important tools for the clean-up of our watersheds in Colorado. While Good Samaritan legislation is needed and would help with some mine clean ups, many more mine sites and watersheds would benefit from the provisions of this bill, especially from the funding this bill provides for clean up by finally placing a royalty on hard rock minerals extracted from public lands.
>
> —*Pete McKay, San Juan county commissioner*

(The) Superfund: CERCLA

For its first fifteen years, Superfund was funded by a federal tax imposed on oil and chemical companies. When the US Congress failed to reinstate the tax after its termination in 1995, the not-so-super trust fund was extinguished by the end of the 2003 fiscal year. Since then, Congress has had to appropriate funds to finance the cleanup at orphaned or abandoned sites with no PRPs. This practice is politically unpopular among many in Congress and has led to protracted debate and budget cuts for the EPA. The result has been a delayed pace of cleanup and restoration activities at many sites. Between 2001 and 2005, a total of eighty NPL Superfund sites were restored and subsequently deleted from the list; from 2006 through 2010, the number of deleted sites fell to thirty-eight; from 2011 through 2015, the number rose slightly to a total of forty-four; from 2016 through July 2018, the number of deleted sites totaled ten ("Deleted National Priorities List Sites" 2017; epa.gov/superfund/deleted-national-priorities-list-npl-sites-state). The listing of new NPL sites, however, has remained relatively constant; from 2001 through 2005, eighty-six new sites were listed; from 2006 through 2010, there were eighty new sites; from 2011 through 2015, eighty-seven new sites were listed on the NPL; and from 2016 through July 2018, thirty-one sites were added to the list, which is lower than the average of sites listed annually (n ≈ 17) during each of the previous five-year periods ("National Priorities List Sites by Listing Date" 2017; epa.gov/superfund/deleted-national-priorities-list-npl-sites-state).

Reinstate the Tax?

In an effort to augment funding and increase the pace of restoration at NPL sites, Congressman Earl Blumenauer (D-OR) introduced H.R. 564—the Superfund Reinvestment Act—in January 2009. It proposed a reinstatement of the Hazardous Substance Superfund financing rate and the corporate environmental income tax on petrochemical industries. Despite having thirty-four cosponsors, the bill never advanced beyond its committee of initial referral, the House Committee on Ways and Means. In the Senate, Frank Lautenberg (D-NJ) introduced similar legislation in the form of S. 3164—the Polluter Pays Restoration Act. It, too, failed to advance beyond its committee of initial referral, the Senate Finance Committee, in the 111th Congress. A year later, the Obama-era EPA took the unusual step of petitioning Congress directly to re-impose the tax, yet it failed to yield any substantive legislative proposals.

Predictably, all efforts were met with stiff resistance from representatives of the nation's petrochemical industries. According to the American Chemistry Council, a trade association representing the largest chemical-making companies, re-imposition of the tax would have been "a lose-lose for the environment and the economy." Further, the association's president and CEO stated in response to the potential reinstatement of the tax: "We paid for sites for which we were responsible, we helped pay for 'orphan' sites where we were not the responsible party, and we paid the corporate environmental income tax. It would be inappropriate and unfair to impose Superfund taxes on companies with no responsibility for site contamination" (Hanson 2010).

The EPA's petitioning of Congress to re-impose the Superfund tax has occurred alongside its funding woes. From 2010 through 2017, the EPA's enacted budget decreased by more than $2.2 billion, and the size of its workforce shrunk by 1,870 employees. Under the Trump administration, more drastic cuts are likely; the president's stated priority was to decrease the EPA's budget by more than 30 percent—from roughly $8.1 billion to somewhere near $5.7 billion for fiscal 2018—and to eliminate up to 25 percent of the agency's current workforce of roughly 15,400 employees. Regarding Superfund in particular, the president proposed a funding cut of 45 percent to enforcement and remediation components—from $404 million in 2017 to $221 million for fiscal 2018 (Tabuchi 2017).

CERCLA also creates potential liabilities for Good Samaritan parties, but in contrast to the CWA, the potential liability for those attempting to restore contaminated sites may take the form of lawsuits for cost recovery—either complete or partial. In particular, CERCLA authorizes the EPA to respond to long-term restoration projects, whereby the agency may require a party to first clean up a contaminated site and then sue for recovery costs from any PRP. The sued PRP may then file a so-called contribution suit against any other PRP. The possible imposition of joint or retroactive liability thus serves as a disincentive for Good Samaritans to engage in cleanup activities at an abandoned site. The only method for parties to protect themselves from this CERCLA liability would be site restoration in full compliance with an official NPDES permit. As previously discussed, CWA compliance often costs millions of dollars, NPDES permitting requirements are onerous, and liability is most likely perpetual, since targeted mines will—quite literally—never stop draining (Churchill 2017).

CONCLUSION

Since the 2015 spill, a temporary water treatment plant near GKM has treated discharge flows between 200 and 800 gallons per minute (gpm), twenty-four hours a day. The treatment plant cost $1.5 million to construct and requires approximately $16,000 per week to operate. It was built by the same EPA-contracted team that originally triggered the 3 million gallon release of AMD into the Animas. The contract provides for forty-two weeks of treatment, with the option to start or stop work as needed. Thus far, the EPA has spent more than $16 million on its response to the GKM spill; since 2015, overall costs have grown to more than $100,000 per day (Marcus 2015).

This chapter examined the GKM blowout and ensuing multi-state spill as a highly visible and potent focusing event. Official designation of the area on Superfund's NPL was formalized after thirteen months, following twenty-five-plus years of fervent opposition by area residents, as well as by local, state, and federal lawmakers. Yet this dramatic policy development was set in motion a mere 110 days after the GKM event when Silverton officials voted to pursue Superfund listing.

The dramatic events that followed the August 2015 blowout centered on the Animas River being turned an unnaturally bright, orange-ish color. The extreme focality of this event confirmed that the discolored river had a central and definitive point of origin—the abandoned GKM high in the Cement Creek drainage. The visual and emotional power of the severely discolored Animas was more impactful than that revealed by numerical indicators associated with water quality and acidity. The symbolic propagation of the orange river through intense international media coverage and public interest had a profound impact on NPL Superfund designation. For these reasons, a more nuanced understanding of the potential impact of focusing events on prompting policy change emerges, for in this case, the deciding factor was the result of GKM's visual and psychic power rather than the scope of ecological damage. The EPA, once seen as a manifestation of the federal government that could not be trusted, was quickly summoned to lead the comprehensive response. With the issue of AMD from abandoned sites elevated on the national agenda and the public's radar, additional policy changes are conceivable, most notably an addition of both royalties and reclamation funds to the 1872 Mining Law, Good Samaritan waivers to the 1972 Clean Water Act, and perhaps even liability waivers for PRPs as defined by CERCLA.

EPILOGUE

In the meantime, an estimated 17 billion to 27 billion gallons of untreated contaminated water will continue to drain each year from forty perpetually draining hardrock mines across the county (Sumi and Gestring 2017). These mines are considered in need of comprehensive, long-term treatment for hundreds to thousands of years, at an estimated cost of between $57 billion and $67 billion annually. Another thirteen mines are considered *likely* to generate water pollution in perpetuity, thus accounting for an additional 3 billion to 4 billion gallons of polluted water each year (Sumi and Gestring 2017). Among these are GKM and three others (Red and Bonita and Sunnyside Mines and the American Tunnel) that currently combine to drain approximately 526 million gallons of AMD annually into Cement Creek.

Finally, the sources of AMD from hardrock mines are not solely abandoned relics of the past. The Pebble Mine has been proposed for construction in an area of south-central Alaska that may contain the second largest deposit of copper, gold, and molybdenum ores in the world, worth an estimated $345 billion to $500 billion (Bluemink 2008). The location of the proposed mine is in the headwaters of Bristol Bay, which has the largest run of wild sockeye salmon in the world; all five species of eastern Pacific salmon spawn in the bay's freshwater tributaries ("Sockeye Salmon" 2017). The Obama-era EPA put a hold on the proposed project in 2014, stating that an operating mine "would result in complete irreversible loss of fish habitat [in some areas of the bay] due to elimination, dewatering, and fragmentation of streams, wetlands, and other aquatic resources" (Griffen et al. 2017). In May 2017, Trump-era EPA administrator Scott Pruitt directed his agency to withdraw the previous protective order for Bristol Bay. During the public comment period, the EPA received more than 750,000 comments in opposition to lifting the order. By late 2017, the developer of the proposed mine had begun the permitting process as part of the mandatory federal environmental review required by the National Environmental Policy Act. If approved, the developer, Pebble Limited Partnership, hopes to enter the construction phase by 2020 ("History/Timeline" 2017). In November 2020, the project was halted when the US Army Corps of Engineers denied a critical CWA permit, stating that proposed waste treatment at the site failed to meet CWA standards ("Army Corps Says No" 2020).

NOTES

1. The *Denver Post* printed 179 articles containing the phrase "Gold King Mine" between 2015 and 2016. A quick search of the *New York Times* online archive yielded 18 articles published between August 7, 2015, and April 3, 2017. The online archives of the *Washington Post* contained 2 articles published between August 10 and August 13, 2015, and the *Los Angeles Times* had 18 articles published between August 8, 2015, and September 9, 2016.

2. Kodas currently serves as associate director of the Center for Environmental Journalism at the University of Colorado in Boulder.

3. On August 5, 2015, news broke revealing the true extent of damage following a ruptured oil pipeline near Santa Barbara, California, when a corroded section of the Plains All American pipeline released an estimated 101,000 gallons of crude into the Pacific. Roughly two months later, an updated figure of 143,000 gallons of crude was disclosed, an amount 40 percent greater than the original estimate.

4. The population of San Juan County, New Mexico, is approximately 130,000.

5. As of March 1, 2016, there were 53 total proposed NPL sites, 1,323 total NPL sites, and 391 deleted NPL sites (US EPA 2016).

6. Since 1970, the League of Conservation Voters has provided factual information about high-profile environmental legislation considered in the US Congress. It produces a biennial National Environmental Scorecard on vote outcomes for bills related to energy, climate change, public health, public lands and wildlife conservation, and spending for environmental programs. See "National Environmental Scorecard" (2017).

REFERENCES

"Army Corps Says No to Massive Gold Mine Proposed near Bristol Bay in Alaska." 2020. SaveBristolBay.org. http://www.savebristolbay.org/in-the-news/2020/11/25/army-corps-says-no-to-massive-gold-mine-proposed-near-bristol-bay-in-alaska.

Baumgartner, Frank R., and Bryan Jones. 1993. *Agendas and Instability in American Politics*. Chicago: University of Chicago Press.

Baumgartner, Frank R., and Bryan D. Jones. 2002. "Positive and Negative Feedback in Politics." In *Policy Dynamics*, edited by Frank R. Baumgartner and Bryan D. Jones, 3–28. Chicago: University of Chicago Press.

Benjamin, Shane. 2015. "Test Fish Removed from Animas River." *Durango Herald*, August 11. https://durangoherald.com/articles/93988-test-fish-removed-from-animas-river.

"Bennet Introduces Bill to Reform Antiquated Hardrock Mining Laws." 2017. US Senate. https://www.bennet.senate.gov/?p=release&id=4045.

Birkland, Thomas A. 1997. *After Disaster: Agenda Setting, Public Policy, and Focusing Events*. Washington, DC: Georgetown University Press.

Birkland, Thomas A. 1998. "Focusing Events, Mobilization, and Agenda Setting." *Journal of Public Policy* 18 (1): 53–74.

Birkland, Thomas A. 2007. *Lessons of Disaster: Policy Change after Catastrophic Events*. Washington, DC: Georgetown University Press.

Bluemink, Elizabeth. 2008. "Pebble's Value Keeps Growing." *Alaska Dispatch News*, February 26. https://www.adn.com/alaska-news/article/pebbles-value-keeps-growing/2008/02/26/.

Butler, Ann. 2015. "Gold King Mine Spill Captures Attention of World's Media." *Durango Herald*, August 21. Print.

Churchill, Ty. 2017. "Understanding the Laws and Regulations Dictating Mine Reclamation." Presentation to the 3rd Meeting of the Citizen Superfund Group, Durango, CO, October 25.

Colten, Craig E., and Peter N. Skinner. 1996. *The Road to Love Canal: Managing Industrial Waste before EPA*. Austin: University of Texas Press.

"Comprehensive Economic Development Strategy." 2015. San Juan County, Colorado. http://www.scan.org/uploads/7__San_Juan_County_Update_2016.pdf.

Congress.gov. 2020. https://www.congress.gov/search?q={%22source%22:%22legislation%22,%22search%22:%22h.r.%207226%22}&searchResultViewType=expanded.

"Deleted National Priorities List (NPL) Sites by Deletion Date." 2017. US Environmental Protection Agency. https://www.epa.gov/superfund/deleted-national-priorities-list-npl-sites-deletion-date.

"Drinking Water Impact Questions." 2015. Colorado Department of Public Health and the Environment. https://www.colorado.gov/pacific/sites/default/files/Drinking-Water-FAQ-09-02-15.pdf.

Finley, Bruce. 2016. "Silverton, San Juan County Vote Yes for Superfund Cleanup of Old Mines." *Denver Post*, February 22. https://www.denverpost.com/2016/02/22/silverton-san-juan-county-vote-yes-for-superfund-cleanup-of-old-mines/.

"Frequent Questions Related to Gold King Mine Response." 2016. US Environmental Protection Agency. https://www.epa.gov/goldkingmine/frequent-questions-related-gold-king-mine-response#impacts.

George, Michael. 2009. "2009 Minerals Yearbook." US Department of the Interior. https://minerals.usgs.gov/minerals/pubs/commodity/gold/myb1-2009-gold.pdf.

"Gold King Mine Spill: How Is the Animas River Faring?" 2016. Mountain Studies Institute. http://www.mountainstudies.org/goldkingspill.

Graham, Edward. 2015. "Updating 19th-Century Mining Law Proves Hard as Rock." *Durango Herald*, November 20. Print.

Griffen, Drew, Scott Bronstein, and John Sutter. 2017. "EPA Head Met with a Mining CEO—and Then Pushed Forward a Controversial Mining Project." *CNN*

Politics, September 22. http://www.cnn.com/2017/09/22/politics/pebble-epa-bristol-bay-invs/index.html.

Hanson, David J. 2010. "EPA Supports Renewal of Superfund Tax." *Chemical and Engineering News*. https://cen.acs.org/articles/88/i26/EPA-Supports-Renewal-Superfund-Tax.html.

"History/Timeline." 2017. Pebble Watch. http://pebblewatch.com/projects/history-timeline/.

Hood, Grace. 2016. "After Years of Opposition, Silverton OKs Superfund Plan." *Colorado Public Radio*. https://www.cpr.org/news/newsbeat/after-years-opposition-silverton-oks-superfund-plan.

Kendall-Ball, Greg. 2015. "Why This Colorado Creek Suddenly Looks Like a Million Gallons of Filthy Yellow Mustard." *Washington* Post, August 7. https://www.washingtonpost.com/news/morning-mix/wp/2015/08/07/clean-up-team-accidentally-sends-1-million-gallons-of-mine-waste-into-colorado-river/?noredirect=on&utm_term=.88f7bf5831fa.

Kingdon, John W. 1984. *Agendas, Alternatives, and Public Policies*, 2nd ed. Boston: Little, Brown.

Kingdon, John W. 2003. *Agendas, Alternatives, and Public Policies*, 3rd ed. New York: Longman.

Langolis, Krista. 2015. "Why Silverton Still Doesn't Want a Superfund Site." *High Country News*, September 3. Print.

Levine, Adeline G. 1982. *Love Canal: Science, Politics, and People*. Washington, DC: Lexington Books.

Limerick, Patricia N., Joseph N. Ryan, Timothy R. Brown, and T. Allan Comp. 2005. *Cleaning Up Abandoned Hardrock Mines in the West: Prospecting for a Better Future*. Boulder: Center for the American West, University of Colorado.

Marcus, Peter. 2015. "A Silver Lining to a Toxic Orange Legacy?" *Coyote Gulch*, October 4. https://coyotegulch.blog/2015/10/04/animasriver-a-silver-lining-to-a-toxic-orange-legacy-environment-america/.

May, Peter J. 1992. "Policy Learning and Failure." *Journal of Public Policy* 12 (4): 331–354.

"National Environmental Scorecard." 2017. League of Conservation Voters. http://scorecard.lcv.org.

"National Pollution Discharge Elimination System." 2017. US Environmental Protection Agency. https://www.epa.gov/npdes/npdes-permit-basics.

"National Priorities List (NPL) Sites by Listing Date." 2017. US Environmental Protection Agency. https://www.epa.gov/superfund/national-priorities-list-npl-sites-listing-date.

"National Priorities List (NPL) Sites by State." 2017. US Environmental Protection Agency. https://www.epa.gov/superfund/national-priorities-list-npl-sites-state#CO.

Paul, Jesse. 2016. "Silverton's Vote on Superfund Letter Won't Happen This Week as Planned." *Denver Post*, January 25. Print.

Pendergast, Art. 2015. "Animas River Disaster: Problems at Mine Site Date Back Decades." *Westword*, August 12. http://www.westword.com/news/animas-river-disaster-problems-at-mine-site-date-back-decades-7021433.

Pratt, Sarah. 2017. "Bringing Geoscience to Bear on the Problem of Abandoned Mines." *Earth Magazine*. https://www.earthmagazine.org/article/bringing-geoscience-bear-problem-abandoned-mines.

"Press Releases." 2017. House Committee on Natural Resources. https://naturalresources.house.gov/newsroom/documentquery.aspx?documentTypeID=1634%3A1639%3A1716%3A1768%3A1866%3A2124&Page=7.

Sabatier, Paul. 1987. "Knowledge, Policy-Oriented Learning, and Policy Change." *Knowledge: Creation, Diffusion, Utilization* 8 (4): 649–692.

Sabatier, Paul. 1991. "Toward Better Theories of the Policy Process." *PS: Political Science and Politics* 24 (2): 144–156.

Schlanger, Zoe. 2016. "The Year in Pollution: Here Are the Nastiest Cases of Toxic Discharge in 2015." *Newsweek*. http://www.newsweek.com/year-pollution-here-are-nastiest-cases-toxic-discharge-2015-410766.

Smith, Conrad. 1992. *Media and Apocalypse: News Coverage of the Yellowstone Forest Fires, Exxon Valdez Oil Spill, and Loma Prieta Earthquake*. Westport, CT: Greenwood.

"Sockeye Salmon (*Oncorhynchus nerka*)." 2017. State of Alaska. http://www.adfg.alaska.gov/index.cfm?adfg=sockeyesalmon.main.

Stone, Deborah. 1997. *Policy Paradox: The Art of Political Decisionmaking*. New York: W. W. Norton.

Sumi, Lisa, and Bonnie Gestring. 2017. "Polluting the Future: How Mining Companies Are Polluting Our Nation's Waters in Perpetuity." Earthworks. https://www.earthworksaction.org/files/publications/PollutingTheFuture-FINAL.pdf.

"Superfund." 2016. US Environmental Protection Agency. https://www.epa.gov/superfund.

"Superfund: National Priorities List (NPL)." 2016. US Environmental Protection Agency. https://www.epa.gov/superfund/superfund-national-priorities-list-npl.

"Superfund Remedial Annual Accomplishments: Fiscal Year 2016 Superfund Remedial Program Accomplishments Report." 2017. US Environmental Protection Agency. https://www.epa.gov/superfund/superfund-remedial-annual-accomplishments#program.

Tabuchi, Hiroko. 2017. "What's at Stake in Trump's Proposed E.P.A. Cuts." *New York Times*, April 10. https://www.nytimes.com/2017/04/10/climate/trump-epa-budget-cuts.html.

"Town of Silverton News." 2016. State of Colorado. https://www.colorado.gov/pacific/townofsilverton/news/town-silverton-news.

Tribune Wire Reports. 2016. "Fearing Stigma, Colorado Contested Superfund Status for Mine." *Chicago Tribune*. http://www.chicagotribune.com/news/nationworld/ct-colorado-mine-waste-leak-20150811-story.html.

US EPA. 2016. "National Priorities List (NPL) by State/US Territory." https://www.epa.gov/fedfac/national-priorities-list-npl-sites-stateus-territory.

Wilkinson, Charles F. 1992. *Crossing the Next Meridian: Land, Water, and the Future of the West*. Washington, DC: Island Press.

4

From Deep Time to Deep Valleys

Hydrology and Ecology of the Animas River Drainage

CYNTHIA E. DOTT, GARY L. GIANNINY, AND DAVID A. GONZALES

The Animas River, like all river systems, is a sum of its parts, providing a unique link among the geography, geology, biology, and climate of the western San Juan Mountains. Many aspects of the Gold King Mine spill are better understood by exploring these interconnections, and this event is an instructive case study for understanding the myriad impacts on river systems, adjacent ecosystems, and the human communities who depend on them (figure 4.1).

The first part of this chapter focuses on the geologic evolution of the western San Juan Mountains, specifically the Bonita Peak Mining District, which includes the Gold King Mine. The geology is especially important in defining the regional landscape, mineralization, human-caused acid mine drainage (AMD) or naturally occurring acid rock drainage (ARD), and buffering capabilities of rocks in the system. As the river flows downstream, so does this chapter, with later portions focusing on where and how the river system interacts with groundwater and floodplain ecosystems in the Animas River Valley. This last part of the chapter focuses on the region where most of society interacts with the Animas River.

DOI: 10.5876/9781646421756.c004

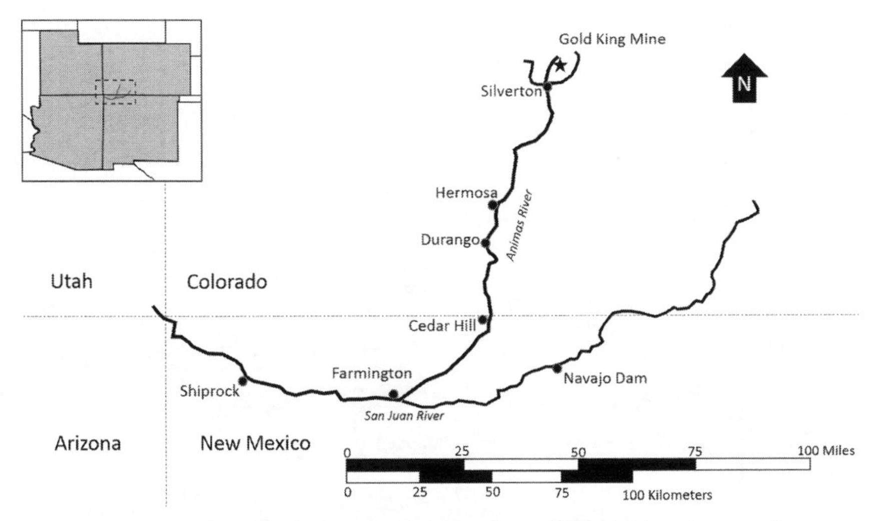

FIGURE 4.1. Map of the Animas River drainage basin from the headwaters above Silverton, Colorado, near the Gold King Mine, to the junction with the San Juan River in Farmington, New Mexico. The San Juan flows in New Mexico through Shiprock and the Four Corners area. In Utah, the San Juan River flows into Lake Powell (not shown).

GEOLOGIC EVOLUTION OF THE ANIMAS RIVER DRAINAGE AND THE SAN JUAN MOUNTAINS

The geologic history is presented as five phases (figure 4.2). The geologic influence on the landscape plays a critical role in the character of the Animas River and helps determine how resilient the river may be to both the Gold King Mine spill and future challenges. In this overview, each phase is shorter in duration than the previous one so it provides more detail on the current landscape.

The current landscape of the Animas River system was significantly shaped by volcanism at ~27 million years ago (Ma) and by glacial and river sculpting during the last 1.5 million years. In both of these times, the area now known as the San Juan Mountains was blanketed by thousands of feet of material: first by volcanic ash and second by the ice of glaciers (e.g., Johnson and Gillam 1995; Johnson et al. 2017; Yager and Bove 2002). Uplift and the combined erosion have shaped valleys and exposed rocks at or

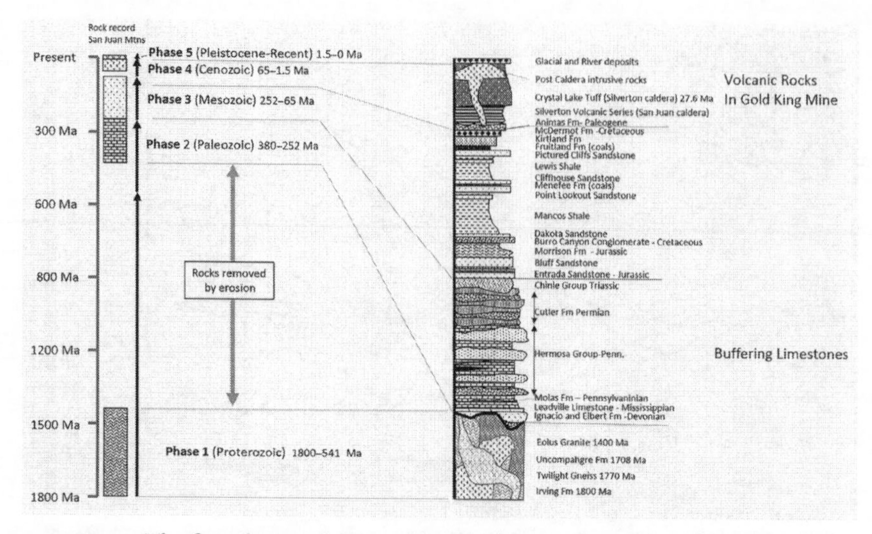

FIGURE 4.2. The five phases of the geologic evolution of the Animas River drainage put in context of a stratigraphic column of the main rock formations in the upper Animas River Basin and the Silverton–San Juan caldera complex. This diagram has time on the vertical axis and thus superimposes the Phase 4 volcanic rocks onto the Animas Formation, which does not occur on the landscape. The spatial relationships of these units are shown in figure 4.4c. Phase 1 rock ages are from Gonzales and Larson (2007) and Karlstrom et al. (2017). Early Phase 3 unit names follow Lucas (2017). Figure modified from Blake (2017).

near the Earth's surface which set the stage for the interaction of streams, groundwater, and mines to create acid mine and acid rock drainage (figures 4.1 and 4.3).

Phase 1 (Precambrian)

Rocks of this phase were created during the assembly of crust on the southern margin of the North American continent and are composed of rocks from ancient plate collisions, similar to those forming in Indonesia today. The oldest exposed rocks in the area are 1,800 to 1,400 Ma (Gonzales and Van Schmus 2007). The first 100 million years of Precambrian history were defined by the formation and accretion of volcanic arcs on the edge of the North American continent. These rocks were metamorphosed in mountain-building events prior to intrusion of granites ~1,700 Ma. Deposition of fluvial

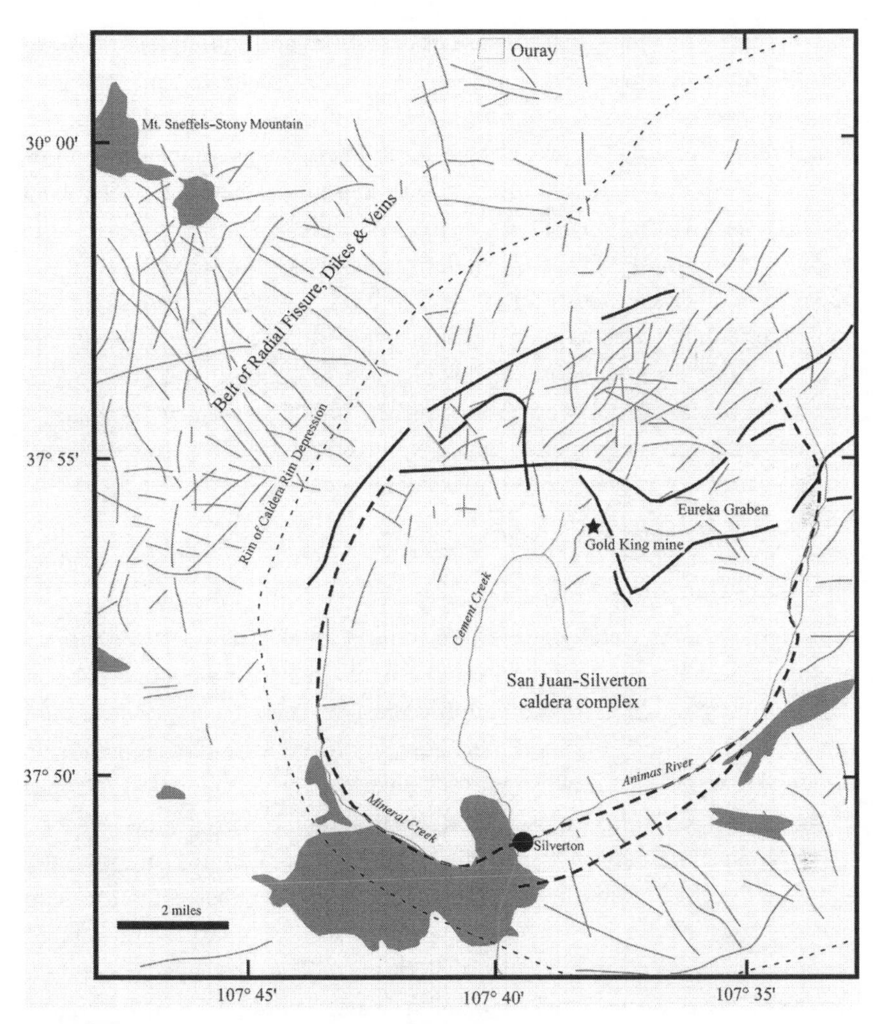

FIGURE 4.3. A generalized map of the San Juan–Silverton caldera complex showing major rivers in the area, location of Gold King Mine, and boundaries of the caldera system. Intrusive masses are shown in gray. Gray lines are fractures, heavy black lines are faults, both of which formed important conduits for mineralizing fluids. The Gold King Mine (shown by the black star) is adjacent to a major fault and fractures. The intrusive mass southwest of Silverton is the ~26 Ma Sultan Mountain stock, one of the post–Silverton caldera intrusions also shown on figure 3.4c (modified from Gonzales and Larson 2017). Intrusive masses around Silverton are often mineralized and were the targets of many mining operations.

and marine sediments (Uncompahgre Formation) and subsequent deformation and metamorphism happened after formation of the volcanic complex. The older rocks were knitted together with the intrusion of younger granites (e.g., Eolus Granite) at approximately 1,400 Ma (Gonzales and Van Schmus 2007). In general, waters that interact with these rocks have a low dissolved metal content (Church et al. 2007). Rocks of Phase 1 are exposed on the southern flank of the Silverton caldera in upper Cunningham Gulch and upper Mineral Creek. Most of the rocks in Phase 1 were metamorphosed or cooled to rock at depths as great as 12 miles (Karlstrom et al. 2017). Before the rocks in Phase 2 were deposited, this region was uplifted, and over 20 kilometers (12 miles) of rock were eroded off the top of these ancient metamorphosed rocks and magma chambers. The missing time along this erosional surface varies from 1,300 to 900 million years (figure 4.2).

Phase 2 (Paleozoic)

Sedimentary rocks were deposited on the erosional surface that developed on the ancient Precambrian rocks and record oceanic and continental conditions from 383 million to 252 million years ago. These Paleozoic strata (figure 4.2) include the Devonian Elbert Formation and Mississippian Leadville Limestone, visible just downstream of Silverton; the prominent limestones of the Hermosa Group exposed at Molas Pass; and the bright red layers of the Permian Cutler Group in the Animas Valley above Durango. Importantly, some of the shallow marine sediments are limestones that contain calcite-rich layers, which can buffer acidic waters and cause metals to be precipitated (Robb 2005). A spectacular example of this is at "Azurite Falls," just downstream of Silverton near the Champion Mine, where acidic AMD waters flow over the 340-million-year-old Leadville Limestone (Luedke and Burbank 2000). This interaction increases the pH (lowers the acidity), causing the copper that was dissolved in the water to be precipitated with carbonate ($-CO_3$), producing a ribbon of the beautiful green mineral malachite ($Cu_2CO_3(OH)_2$) along the falls.

Phase 2 culminated in another plate collision (which was caused by the assembly of the super-continent Pangaea), which led to the formation of the Ancestral Rockies, a tropical mountain range that was completely eroded away by the early portion of Phase 3 (Kluth and Coney 1981; Leary et al.

2017). Surface waters from drainages with the Paleozoic strata are typically high in dissolved calcium (Ca^{2+}) that can also reduce the acidity of streams. In Cement Creek and Mineral Creek in the Silverton caldera, the yellowish-orange iron hydroxide coating in streams emanating from mines or mine tailings typically has a pH of 2.0 to 4.5. Where these streams mix with the calcium-rich waters of other streams, pH rises to less acidic conditions (pH 5 to 6), and white aluminum hydroxide precipitate coats the stream beds. In high-volume, low pH streams like Cement Creek, there are no adequate calcium-rich waters available to significantly increase the pH until it is diluted with the large volume of water in the main stem of the Animas River (Church et al. 2007).

Phase 3 (Mesozoic)

This period of geologic history is defined by additional plate collisions to the west that caused the Earth's crust to subside in the Four Corners area, making it a prime location for sediment to accumulate. A thick set of sedimentary rocks (figure 4.2) were deposited—first by meandering rivers with red (iron-rich) sediment (Chinle Group) (Lucas 2017), then as dune fields (Entrada and Bluff Sandstones) (Lucas 2017), and next as the river and floodplain sediments that formed low-angle aprons of sediment from mountain ranges in central Utah (Morrison and Burro Canyon Formations). Finally, the top of the stack of Phase 3 sediments includes shoreline sandstones and deeper water shales of an inland sea (Dakota Sandstone, Mancos Shale, Mesa Verde Group, Lewis Shale, Pictured Cliffs Sandstone, Fruitland Formation, Kirtland Formation). These tan sandstones and gray/black shales make up the rock layers in and around Durango.

During the 210-million-year-long Phase 3, western North America was reshaped by the collision of plate fragments on the western margin and subduction of the Farallon oceanic plate. Near the end of Phase 3, compression caused uplift and tilting of the strata in the Animas drainage, creating folding that arched sediments across the San Juan dome (figure 4.4). This inclined the strata and the Earth's surface to the south from Silverton and to the north from Telluride, Ouray, and Ridgeway. This tilting was the birth of the south-flowing drainages of the San Juan dome, including what would become the Animas River (Cather 2004).

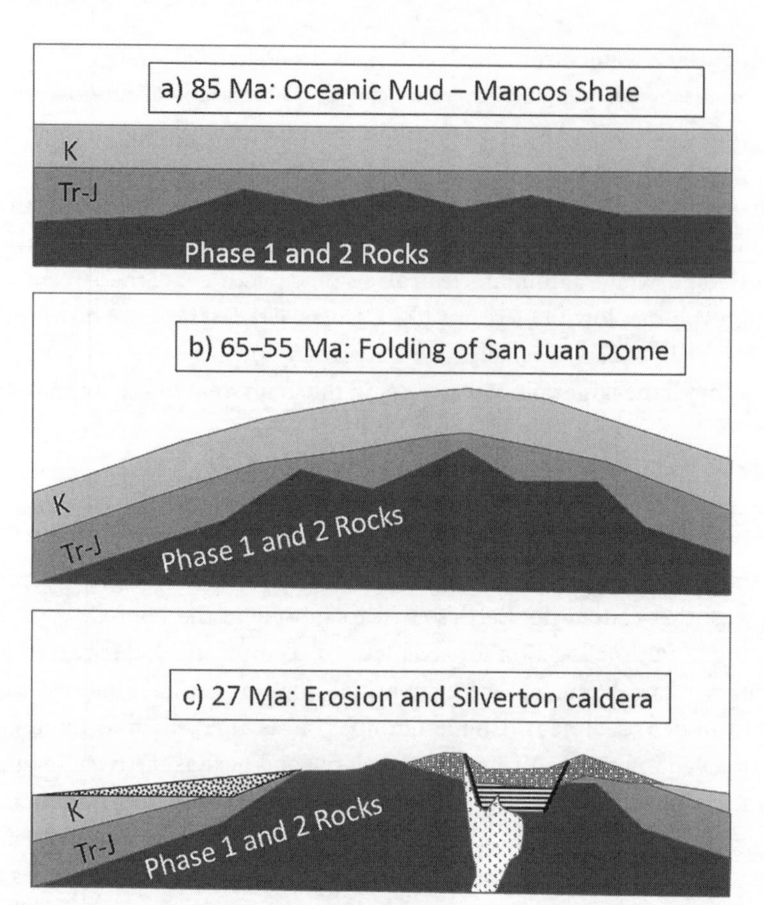

FIGURE 4.4. A schematic diagram showing a three-step evolution of the San Juan Mountains north of Durango: (a) 85 million years ago, flat-lying ocean-bottom muds (Mancos Shale) covered the entire region. They buried the worn-down remnants of the Ancestral Rockies and all previous rocks. (b) 75 million to 60 million years ago, the San Juan dome was formed by a combination of compression and injected magmas. The tilted strata on the south side of the San Juan dome initiated south-flowing drainages that would eventually include the Animas River. (c) post-60 Ma the San Juan dome had been eroded, beveling Phase 3 strata of the Triassic, Jurassic (Tr-J), and Cretaceous (K) geologic periods and producing the conglomerates of the 65 Ma to 50 Ma Animas Formation (Pecha et al. 2018) ("A" on the diagram). The San Juan and Uncompahgre calderas erupted 28.5 million to 28 million years ago, depositing unit "B" with the Sapinero Tuff and Silverton Volcanic Series. At 27.6 Ma, the Crystal Lake Tuff (unit "C") was erupted, causing the collapse of the Silverton caldera. After the Silverton caldera collapse, younger magmas and mineralizing fluids (unit "D") followed faults and fractures. *Courtesy*, G. L. Gianniny.

Phase 4 (Cenozoic)

At about 75 million to 65 million years ago (Gonzales 2015), plate tectonic collisional events on the western margin of North America triggered the injections of magma into the region. Evidence of this igneous activity lies in the eroded remains of the first volcanoes, laccoliths (e.g., La Plata Mountains), and some localized hydrothermal alteration and mineralization in Rico and Ouray (as summarized in Burbank and Luedke 2008; McKnight 1974). Currently, there is no evidence of significant alteration or mineralization in what would become the Animas drainage in Phase 3. Both tectonic compression and magmatic inflation caused uplift in and subsequent erosion of the San Juan Mountain region after 60 Ma, along with deposition of river deposits at ~65 Ma (Pecha et al. 2018) (figure 4.4b). Although the record is sparse, it appears that at the end of Phase 3, the crest of the San Juan dome experienced erosion for nearly 30 million years (Donahue et al. 2012). This erosion removed most of the Phase 2 and 3 (Paleozoic and Mesozoic) rocks from the top of the San Juan dome, leaving a dissected landscape defined by tilted Phase 1 and early Phase 2 rocks (figure 4.4c) that were about to be buried by lava flows and thousands of feet of volcanic ash.

Time of Volcanism and Mineralization

The San Juan Mountains were the site of numerous volcanic super-eruptions from around 37 million to 23 million years ago (Lipman 2006; Lipman and McIntosh 2008; Lipman et al. 2015). This period of geologic history is defined by at least twenty-eight major eruptions with individual volumes of 100 to 5,000 km^3 from collapsed calderas (similar to Yellowstone) in southwestern Colorado and northern New Mexico (Lipman 2007; Lipman and Bachmann 2015). In addition to caldera-type eruptions, there was also widespread eruption of andesitic lavas and lesser pyroclastic rocks from stratovolcanoes and related vents (similar to those in the Pacific Northwest).

It is nothing short of terrifying and awe-inspiring to imagine the scale of the volcanic eruptions that occurred in the San Juan Volcanic Field. The largest ash eruptions caused the volcanic region north of present-day Creede to collapse to form the massive La Garita caldera, which was 35 km × 100 km across. This event erupted a blanket of ash called the Fish Canyon Tuff, which was so hot it re-melted (welded) after falling from the sky. This blanket of tuff has a volume of at least 5,000 km^3, which is enough to bury the entire state of Colorado in 18.5 m (60 ft.) of ash (Lipman 2006). For perspective with other volcanic eruptions, in 1980, Mount Saint Helens erupted 1.6 km^3 of ash; and the largest of the Yellowstone "super-volcano" eruptions, 2.1 million years ago, spewed approximately 2,500 km^3 of ash, which landed over much of North America (Rivera et al. 2014).

Caldera eruptions in southwestern Colorado started ~28.5 Ma with the simultaneous collapse of two connected calderas (San Juan–Uncompahgre complex) during eruption of 1,000 km^3 of eruptive material that comprises the ~28.5 Ma Sapinero Mesa Tuff (Lipman et al. 1973; Luedke and Burbank 2000; Steven and Lipman 1976) (figures 4.2 and 4.4c). The depression formed by the San Juan caldera and its flanks were covered and filled by approximately 1,000 m (3,000 ft.) of post-caldera lavas and sediments of the Silverton Volcanic Series (Lipman et al. 1973; Luedke and Burbank 1969, 2000). This volcanic pile includes the lower Burns Member lavas and tuffs, an upper Pyroxene Andesite Member, and the Henson Formation. Where these rocks are not mineralized, they contain calcite, epidote, and chlorite, which can buffer acidic water.

The Silverton caldera was formed by another period of collapse and violent eruptions at ~27.6 Ma during eruption of the Crystal Lake Tuff (figure 4.4c). The 15 km wide Silverton caldera is nested within the San Juan caldera (figure 4.3), with an estimated volume of 25 to 100 km^3 (Lipman 1996, 2006). Although it is roughly circular, the collapse was asymmetric and likened to a "trap door" with a sharp faulted margin in the southeast and layers of volcanic rock dropped into the caldera. The formation of the San Juan–Silverton calderas was preceded by inflation of the region as magma moved to higher levels. This broad uplift and subsequent collapse created numerous fractures (Luedke and Burbank 2000) parallel to caldera walls (concentric) extending out from the central collapse like spokes on a wheel (radial fractures) (figure 4.2). For 20 million years after the collapse of the Silverton caldera, the area was re-inflated and infiltrated by small bodies of magma, which caused further fracturing. These fractures created pathways for the "plumbing" system that controlled the movement of mineral-charged hydrothermal waters that invaded the area (Gonzales 2015; Yager and Bove 2002) and also influenced the movement of water at the surface and in the subsurface today.

Phase 4 events thus proved to be a critically important stage in the evolution of the San Juan Mountains and the Bonita Peak Mining District in particular. Volcanic rocks created by the San Juan and Silverton calderas in the western San Juan Mountains host the Gold King Mine and the entire Bonita Peak Mining District (Yager and Bove 2002). The magmatic activity and deformation of this area from 28 million to 5 million years ago created the sources of metals, fluid plumbing, and hydrothermal systems that set the stage for the formation of mineral deposits by creating pathways for fluids containing dissolved metals. The precipitation of minerals in veins and localized replacement deposits produced sulfide minerals such as pyrite that have contributed to acidic waters ever since they formed. Although the timing of mineralization throughout the region is not fully constrained, there is evidence that within the caldera systems it happened from 15 million to 5 million years ago (e.g., Casadevall and Ohmoto 1977; Fisher 1990; Gonzales 2015; Lipman et al. 1976; Yager and Bove 2002). This mineralization led to the vast mineral wealth of the San Juan Mountains and also created potential for the mobilization of acidic, heavy metal–laden waters we are experiencing today.

Phase 5 (Pleistocene to recent)

On and off glaciation occurred over the last 1.5 million years but could have started as late as 0.8 million years ago, with the most recent glaciation occurring from about 21,000 to 14,000 years ago (Guido 2007; Johnson and Gillam 1995; Johnson et al. 2017). Two main influences of glaciation are key in framing the story of the Gold King Mine spill impacts. First, glaciers sculpted peaks (e.g., Bonita Peak), exposing the rocks formed in the magmatic events after 28 Ma, and they carved deep U-shaped valleys like the Animas River Valley north of Durango. Second, glaciers moved large volumes of eroded sediment. The ice itself transports sediment down-valley like a conveyor belt, and the ice also spawns rivers that carry huge volumes of sediment, especially during the glacial melt phase. Many of the sand and gravel deposits that drape the landscape from Durango to Farmington were generated by glaciers and glacial rivers (Gillam 1998). Thus the combination of ice, water, and transported sediment shapes drainage basins like the Animas River watershed, leading to variations in stream gradient and valley width in the present landscape. These sediments are also important for storing groundwater within the river valley.

Acid Rock and Acid Mine Drainage

The youngest part of Phase 5 includes the history of mining (see chapter 1, this volume) and its impacts that caused acid mine drainage. Studies of the geochemistry and paleontology of pre-mining sediments in the Animas drainage indicate that heavy metal loading in the Animas River increased twofold to eight-fold after mining (Church et al. 2000). As described in the introduction (Clark, this volume), AMD and naturally occurring ARD are caused by the dissolution and oxidation of sulfide minerals like pyrite (iron sulfide, or FeS_2) in a cascade of reactions that involve both inorganic and organic processes. The outcome is low pH, acidic waters created by the interaction of sulfur and hydrogen (to form sulfuric acid), which can dissolve minerals and release heavy metals into water systems. Other sulfide minerals are also important in this process as they contribute not only sulfur but also, in low pH conditions, dissolved metals. Sphalerite (ZnS) contributes zinc, which in moderate concentrations is deadly to many fish; galena (PbS) yields lead, which causes developmental and nervous system disorders in vertebrates (Sfakianakis et al. 2015); chalcopyrite (Fe, CuS_2) is a source of both iron and copper, and tetrahedrite ($(Cu, Fe)_{12}As_4S_{13}$) and enargite (Cu_3AsS_4) are potential sources of arsenic. The abundance of pyrite and chalcopyrite in the mineralized areas creates a large surplus of iron in the system, which is deposited as a coating of yellow-orange iron hydroxide in AMD- and ARD-impacted streams. This iron hydroxide coating combined with the low pH and heavy metal concentrations prevents photosynthetic algae and bacteria from developing a base for a photosynthetic food chain. In addition, micro- and macro-invertebrates do not survive in these toxic waters, and neither can their predators like fish. Thus AMD and ARD streams are sterile with the exception of acid-loving bacteria.

The five-phase evolution of the western San Juan Mountains described in the first section of this chapter illuminates the origin of the rocks, minerals, acid drainage, and sediments, as well as the formation of the Animas River watershed. In the second portion of this chapter, we shift our focus to three tightly interconnected facets of the Animas River system: the specific shaping of the landscape (geomorphology) and how this sets trajectories first for hydrologic systems and then for the superimposed ecological systems of floodplains.

GEOMORPHOLOGY OF RIVER SYSTEMS

River systems are studied at many spatial scales, from the full drainage basin (watershed) to sub-segments of the full river called "reaches." A river reach is a segment of the river channel and its surrounding banks and floodplain that shares the same geomorphologic setting such as consistent gradient, sediment type, and valley shape (Gregory et al. 1991). One aspect of river reaches relates to processes of sediment movement. Some regions are dominated by sediment erosion, others by sediment transport, and still others by deposition and accumulation. In mountain drainage systems like the Animas River, headwater streams are often steep and narrow, carved into deep valleys, and are dominated by processes of sediment erosion. Further downstream, a pattern of alternating zones of sediment transport and sediment deposition is established, depending on reach-scale variations in stream gradient. From the air, this pattern may look like "beads on a string" (Ward et al. 2002), with higher-gradient reaches of sediment erosion and transport punctuated by wider "beads" with low-gradient, sediment accumulation and floodplain development (figure 4.5). It is the broad floodplain reaches that are of greatest concern for potential contamination from Gold King spill dissolved metals. Ironically, these are the same zones where many human activities such as farming occur, since the soils are often finer-grained and nutrient-rich. Agricultural activities including orchards, annual crops, hay production, and livestock grazing often occur in floodplain settings. In more recent years, wider floodplain valleys have often been favored sites for residential development in suburban or exurban areas. In most cases, people living outside of metropolitan areas in the West get their water from wells, which can have an intimate connection with the river that formed the floodplain.

HYDROLOGY OF RIVER SYSTEMS

Rivers move water from their headwaters to their mouth, or junction with another stream. In the case of the Animas River, the headwaters are high in the San Juan Mountains above Silverton, and the mouth is where the Animas flows into the San Juan River in Farmington, New Mexico (see figure 4.1). While people tend to think of rivers as transporting only the water on the surface, a large portion of the water held in a river system is stored and moved as groundwater beneath the surface, which is critical for

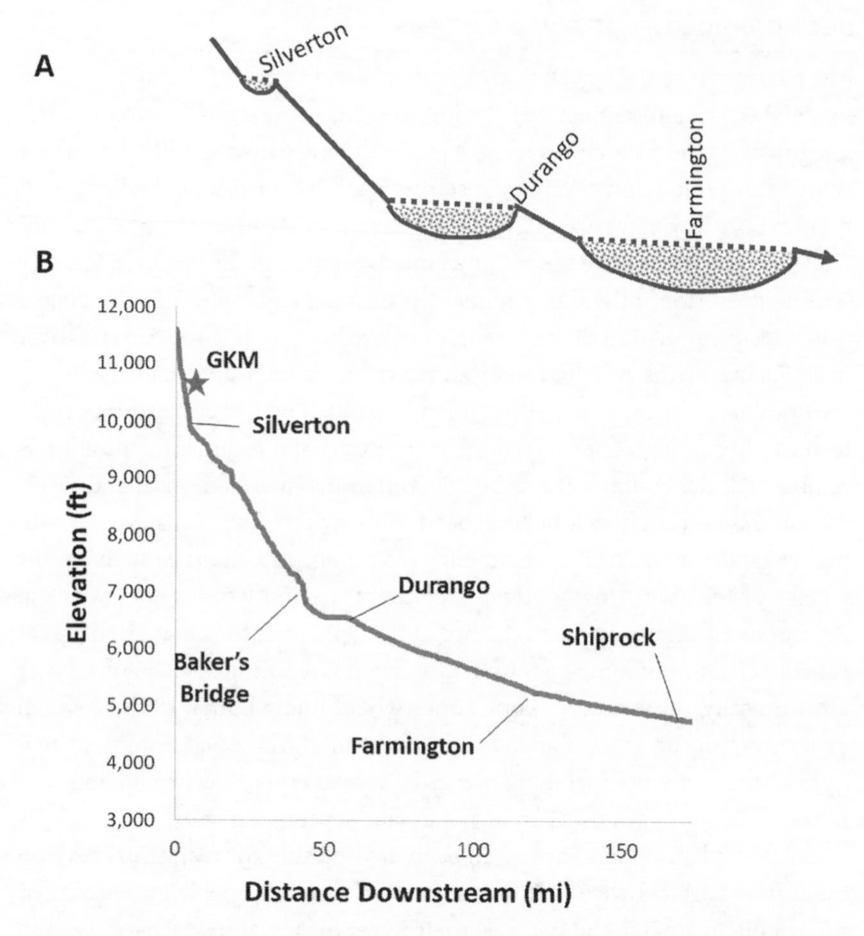

FIGURE 4.5. "Beads on a string" conceptual model of floodplain development (after Ward et al. 2002): (a) schematic profile of the Animas River showing steep, high-gradient reaches and lower-gradient reaches with sediment accumulation and floodplain development in the Silverton valley, the valley north of Durango, and the reaches upstream and downstream of Farmington, New Mexico; (b) longitudinal profile of the Animas River showing areas of steep versus low gradient. Gradient data compiled by J. Harvey.

plant and animal life and plays a key role in dissolved substance transport. Indeed, surface water and groundwater act as a single, integrated system (Winter et al. 1998), with many fascinating and complex interactions occurring between them.

Surface Water

Mountain rivers like the Animas have a seasonal pattern of discharge (measured in cubic feet per second, or CFS) that is dominated by snowmelt. The highest flows in most years occur during snowmelt in the headwater tributaries, which historically spanned from May through June, tapering off in July (figure 4.6a). These snowmelt peaks typically range from 4,000 CFS to 6,000 CFS in Durango, where the longest-recording US Geological Survey (USGS) stream gauge on the Animas is located (USGS 1912–present). The highest spring peak in the last fifteen years occurred in May 2005, at 8,070 CFS. In recent warmer years, the snowmelt peak has started earlier, sometimes as soon as March, and can end by late May. Discharge tapers down during the summer months, but in our region the North American Monsoon delivers water in the form of convective thunderstorms during mid-July to early September and may cause short-lived spikes in river discharge (figure 4.6a) (Hereford et al. 2002; Webb et al. 2007). These sudden high-flow events from tributaries often deliver high pulses of sediment to the river channel (Webb et al. 2007). Fall storms are rare, but tropical storm remnants from either the Gulf of Mexico or the eastern North Pacific occasionally deliver heavy rains to the region; when combined with low-pressure systems that help lift the moist tropical air over some portion of the Southwest, large floods can potentially result (Hereford and Webb 1992; Smith 1986; Webb et al. 2007). The largest recorded flows on the Animas River (25,000 CFS in Durango) occurred on October 5, 1911, due to the rare confluence of a tropical cyclone off the Pacific pumping moisture into a stalled low-pressure system over the Utah-Arizona border (Smith 1986; Webb et al. 2007). Rainfall recorded in Gladstone, Colorado, a mining town above Silverton adjacent to the Gold King Mine, was 8.05 inches on October 5 (Smith 1986). In other words, about one-fourth of the annual precipitation total was delivered in one day. In most years, river discharge gradually drops throughout the summer and fall; during the winter months low or "base flow" (commonly close to 200 CFS on the Animas in Durango) is maintained by groundwater percolating back into the river channel from the surrounding floodplains and uplands.

Where surface flows in stream channels are concerned, dilution is indeed a solution to pollution. When the flow of water in the channel is low, any input of sediment, dissolved metals, effluent, or excess fertilizer will be concentrated and is likely to have a larger impact on life in the channel than it

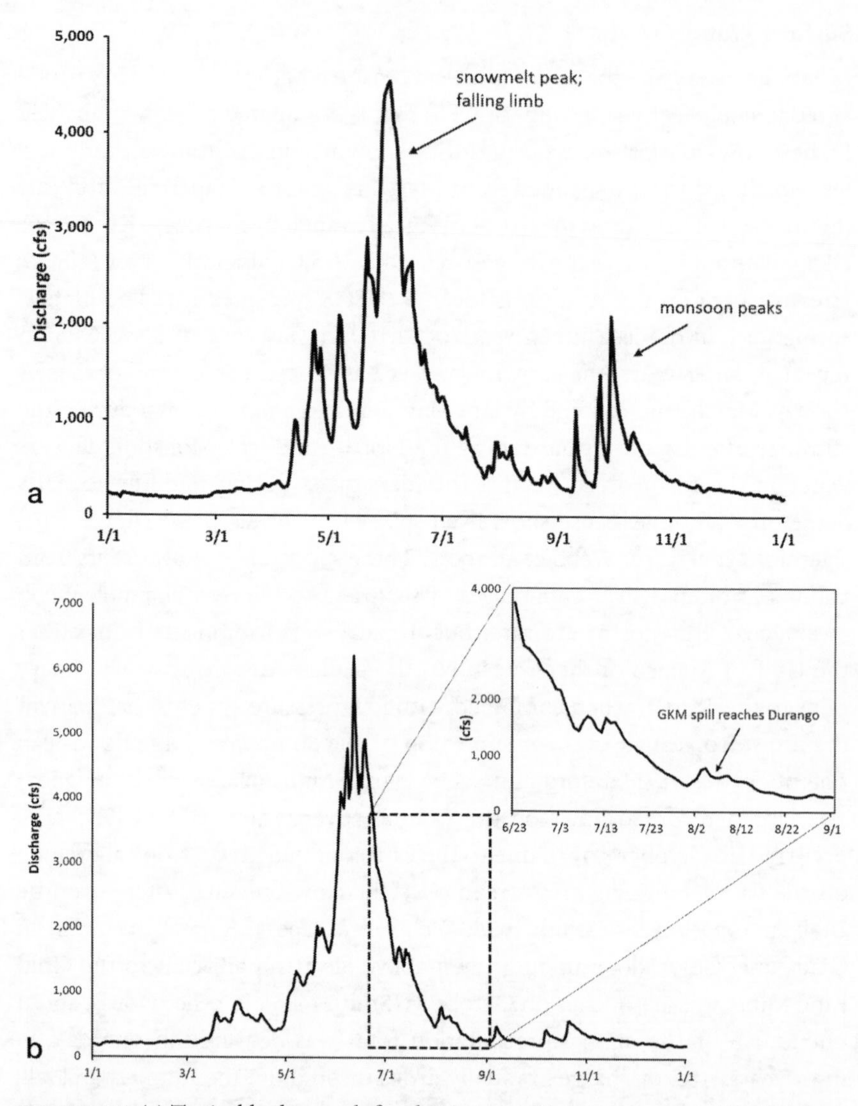

FIGURE 4.6. (a) Typical hydrograph for the Animas River at Durango (USGS gauge 09361500). Data from the 2014 calendar year show a high early summer peak due to snowmelt in the mountains and distinct late summer to fall peaks due to monsoon rain events. (b) 2015 hydrograph for the Animas River at Durango. The Gold King Mine spill arrived in Durango on August 6 (see inset graph), on the falling limb of the hydrograph, in a year with only weak monsoon flow peaks.

would if it occurred during a period of high flow volume. The Gold King spill is an excellent case in point. Although 2015 began with a dry winter and an early spring, it was a wet early summer, including "Miracle May" as dubbed by water managers in the region, which delivered record rainfall for the State of Colorado (WWA 2015). The snowmelt peak discharge was 6,730 CFS on June 11 (figure 4.6b). On August 6, the day of the Gold King spill, the river was well into its summer decline in flow (called the "falling limb" of a hydrograph), punctuated by occasional monsoon flow spikes. The flow in Durango on August 6, before the spill arrived, was fluctuating around 600 CFS. Had the spill occurred during the spring runoff, the sediment and dissolved metals it carried would have flushed through both the Animas and San Juan River systems, and little of that sediment would probably have been deposited along the rivers' banks. But because the spill occurred in the summer, during the falling limb of the hydrograph, the slowing flows of the rivers meant that much of the Gold King sediment was deposited along their banks. As was illustrated earlier, most sediment deposition occurs in lower-gradient river reaches with wider valleys. The first such valley along the Animas River below Silverton is the broad glacial valley of the meandering Animas north of Durango, which is followed downstream by many more such broad reaches that host abundant agriculture.

Groundwater

Water below the surface of the land is found in two zones: the *unsaturated zone*, where the pores between grains of sediment are filled with a combination of air and water, and the *saturated zone*, where all of those openings are filled with water. This saturated zone is the area known as groundwater, and the upper surface of this zone is called the *water table* (Webb et al. 2007; Winter et al. 1998). A continuous saturated zone is called an *aquifer*, and in wide river valleys this is often referred to as the *floodplain aquifer*. Most plant roots grow in the unsaturated zone because of their need for oxygen, though some plants that grow in river floodplains are adapted to have their roots penetrate below the water table and into the saturated zone for at least a part of the growing season. Also of note, water wells are drilled into the saturated zone, which in a floodplain setting is typically shallow and interacts with the adjacent river channel.

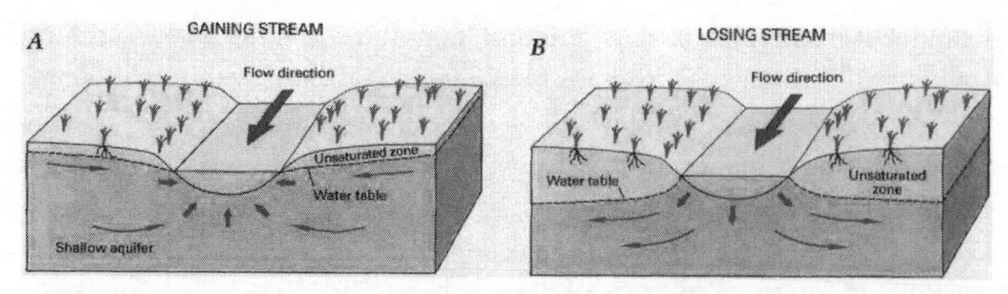

FIGURE 4.7. Diagram showing gaining versus losing stream conditions (from Winter et al. 1998, figure 8: https://pubs.er.usgs.gov/publication/cir1139). In a gaining stream (a), water flows from the surrounding floodplain and upland sediments into the river channel. In a losing stream (b), the net flow direction of water is from the river channel out into adjacent floodplain alluvial sediments.

Groundwater *recharge* occurs through a combination of precipitation, runoff from higher land, and interchange with surface water bodies. This means that the surface water of a river corridor is intimately connected with its groundwater, in such a way that the water in the river channel extends both downward (if the streambed is permeable rather than bedrock) and outward, laterally into the adjacent floodplain sediments (figure 4.5). Thus just as the surface flows in the river channel are dynamic and subject to seasonal changes, the water table in the floodplain is also variable and can fluctuate up and down with the seasons and between years (Dott et al. 2016; Winter et al. 1998). It is the "beads" of the wide floodplain valleys with a higher volume of sediment deposits that have the most dynamic interactions between the floodplain aquifer and surface flow in the river channel. In these broad valley reaches, groundwater may flow into the river from the floodplain in some zones, in which case the river is called a *gaining stream*. In other cases, the water from the stream channel may flow into the adjacent floodplain aquifer, creating a *losing stream* (figure 4.7) (Winter et al. 1998). Many reaches of rivers switch between gaining and losing with the seasons: gaining water from the floodplain in the spring after the floodplain aquifer has been recharged by local and regional snowmelt and losing water to the floodplain later in summer when the water table has dropped to a level below the water surface in the stream channel (figure 4.7). The potential influx of dissolved metals into the adjacent floodplain aquifer is most significant in reaches with floodplain development and during seasons dominated by losing stream conditions. However, in some cases, the floodplain sediments can act as a natural cleanser of contaminated waters by causing adsorption of metals and precipitation of minerals (Maltby and Acreman 2011).

In the Animas River Valley, little was known about groundwater dynamics prior to the Gold King spill. Extensive study of existing and new water wells in the New Mexico portion of the floodplain indicates that most of the river reaches in this region exhibit gaining stream conditions throughout the year, with the exception of an area around Cedar Hill and Aztec where losing stream conditions could exist during summer and winter (Blake 2017; Timmons et al. 2016). Conditions in the Colorado portion of the floodplain are unknown. Many of the wells sampled in the alluvial aquifer of the Animas valley did exhibit high levels of some metals, including manganese, iron, and aluminum; but these probably indicate many years of accumulated impacts on water quality as opposed to a distinct signal from the Gold King spill (Blake 2017; Timmons et al. 2016; see also EPA 2018; SJBPH 2018).

The scale of variation from localized gaining to losing stream segments can vary from 10 meters to kilometers. In highly interactive zones, this is referred to as "hyporheic flow" (Magliozzi et al. 2017). This is one more location in the floodplain where metal-laden waters can interact with floodplain sediments. An additional complexity in floodplain aquifers is the impact of unlined irrigation ditches and irrigation in general, which provides an additional source of groundwater recharge and potential mixing with river waters.

FLOODPLAIN ECOSYSTEMS

The landforms described in this chapter so far truly set the stage for the biological communities that inhabit the region. Because the headwaters of the Animas River are above 12,000 feet in elevation, it flows through several ecological life zones. The river starts in the alpine tundra and continues through subalpine and montane forests (above Durango) all the way down into high desert pinyon-juniper woodlands (see figure 4.1) (at 5,000 feet elevation), where it joins the San Juan River in Farmington, New Mexico (Spencer and Romme 1996).

Along the river corridor itself, specialized riparian (or "river's edge") plant communities develop depending on both elevation and valley configuration. In areas with steep gradients and narrow valleys (the "string" portion of the beads on a string model, figure 4.5), vegetation from the surrounding uplands will usually extend down to the river's edge, with a narrow band of deciduous shrubs like willows and alders growing right by the water. In broader valleys with lower stream gradients and more sediment accumulation (the "beads" of the model), true floodplain habitat may develop with more diverse vegetation that provides important wildlife habitat. In the Silverton valley (at and above 9,000 ft.), this riparian zone has been highly disturbed by mining activities, and the prevalence of mine tailings limits the development of plant growth. However, there are still areas where high-elevation riparian habitat may be found, both in the main Animas valley around and above Silverton and up tributaries like Mineral Creek and Cunningham Gulch. Subalpine wetlands often develop in such valleys, with extensive thickets of willows (usually many species growing together in what's called a "willow carr"); dense growths of sedges, rushes, and grasses; and scattered stands of alder, river birch, or other deciduous shrubs. These areas are often highly engineered by beavers, which create ponds and extend the width of the wetland, providing important habitat for fish as well as for moose, elk, and a high diversity of songbirds (Somers and Floyd-Hanna 1996).

In the middle elevations (6,000 to 8,000 ft.), similar broad valleys in the montane life zone have a more complex structure that includes both an upper canopy of trees and the lower canopy of shrubs. The Animas River Valley above Durango has well-developed riparian habitat and so-called cottonwood galleries: large stands of cottonwood trees with a rim of willows growing along the river's banks. Much of this floodplain has also been developed for agriculture and is used for livestock grazing, but in many areas the cottonwood-willow woodlands are still fairly intact. North of Durango, the cottonwood-willow woodlands are dominated by narrow-leaf cottonwood (*Populus angustifolia*), which has an elevation range of 5,000 to 10,000 ft. However, the lower-elevation, broad-leaved plains cottonwood (*Populus deltoides*, range of 3,500 ft. to ~9,000 ft.) can also be found in this region, and the two species' ranges overlap right in Durango (Culver and Lemly 2013). Where they overlap, the species hybridize to create lance-leaf cottonwood (*Populus x acuminata*). South of Durango, the cottonwood woodlands are dominated by the plains cottonwood.

The typical structure of a riparian woodland in this life zone consists of a band of trees covering much of the floodplain, with a diverse understory of shrubs and herbaceous (non-woody) plants and then a band of dense willows right along the water's edge. Colorado is home to over forty species of willows (Kittel 2016), but the most common and abundant in this elevation zone is coyote or sandbar willow (*Salix exigua*). Other willow species are often interspersed in these thickets, along with red osier dogwood (*Cornus sericea*) and silverleaf buffalo berry (*Shepherdia angustifolia*). As with the higher-elevation riparian zones, these areas in mid-elevation are important for wildlife habitat, as they provide water, forage, and cover even for species that normally live in upland habitats (Culver and Lemly 2013; Naiman et al. 1993; Nilsson and Berggren 2000).

Floodplain habitats play an important role not only for wildlife and biodiversity but also because of the myriad ecosystem services they provide. These are services the natural system supplies at no cost to humanity, but to great benefit. Key features provided by floodplains include the reduction of downstream flood risk and, most important to this discussion, both aquifer recharge and groundwater filtration, which often results in improved water quality (Maltby and Acreman 2011). Thus it is useful in considering the potential fate of AMD runoff to understand the dynamics of floodplain hydrology and vegetation, as they may play a role in determining the fate of heavy metals in the system.

An important dynamic is established on the floodplain between the bands of vegetation and the floodplain aquifer. Because the plants that grow in floodplain habitats are adapted to high water use, they rely on a shallow water table to support their growth. Mature cottonwood trees tend to grow with their major roots extending down to the top of the water table but not into it (Rood et al. 2011; Williams and Cooper 2005). Of course, since the level of the water table fluctuates with the seasons and also between years, sometimes the roots will be inundated as the water table rises, and other times they will be left in drier soils above the top of the water table. But in general, cottonwood rooting depth marks the average level of the water table. Interestingly, this depth appears to be established when the tree is young, probably in its first year of growth. As its roots chase the declining water table in the summer, they may grow as fast as 0.5 to 1 cm/day (Mahoney and Rood 1998) until base level is reached and the roots stop in the zone right above the water table. Here they can uptake large volumes of water through capillary action without being subjected to the anoxic conditions present below the water table.

In contrast, many willows, including the coyote willow (*Salix exigua*), grow best when their roots extend below the water table and down into the groundwater (Amlin and Rood 2002). This explains the zonation seen in riparian woodlands, with the willows right along the river's bank. During high-water periods, willows often have their roots growing out into the river channel, and new shoots of coyote willow are found even in late summer growing out to the river's edge, since this plant can spread by underground runners (rhizomes) where enough water is present (Douhovnikoff et al. 2005).

The implications of the Gold King Mine spill for plant growth are naturally concerning but also interesting. Unfortunately, there are few pre-spill studies that would allow one to compare plant responses to the Gold King spill pulse of heavy metals in the Animas River. However, it is possible to compare the heavy metal content of riparian and wetland plant roots and leaves in the Animas River with those of the Florida River, which typically has far lower concentrations of dissolved heavy metals. Magena Marzonie, an undergraduate biology major at Fort Lewis College, in collaboration with Dr. Callie Cole (Department of Chemistry) and Dr. Cynthia Dott (Department of Biology), investigated this and helped elucidate some of the characteristic differences of metal accumulation in Animas River vegetation (specifically *Salix exigua*) (Marzonie 2017). Sampling in 2016, a year after the spill, Marzonie found that the root samples from the Animas River almost always had higher concentrations of heavy metals (especially zinc, cadmium, manganese, and iron) than did roots from the Florida. A lower-gradient site just above Durango typically had the highest concentrations of most metals. Most of the leaf samples did not have significantly higher metal loads than the control site, suggesting that metal transduction from roots to leaves, at least in willows, tends not to occur. This is an encouraging result for agricultural users of the floodplain but would need to be replicated.

Finally, a great concern when any pollutant is introduced into a water body is what the impact will be on aquatic organisms like fish and their invertebrate prey. This is partly because aquatic organisms can be very sensitive to water quality, and some species or groups may be completely eliminated from the system at even moderate levels of heavy metal contamination (Clements et al. 2000; Sfakiankis et al. 2015). Fish are impacted by heavy metals in several ways, especially during development. Developmental deformities of the spine are common in young fish and impact their ability to navigate, find food, and escape predation (Sfakiankis et al. 2015). In addition, depending on the type and form of heavy metals present, fish metabolic processes and physiological systems may be impaired, often involving impacts to gill function (Sfakiankis et al. 2015). In the case of the Gold King spill, apparently the metal-laden water flushed through quickly enough to have minimal impacts on fish abundance and health, at least over the short term (White 2016). Fish tissue samples were analyzed in both August 2015 and March 2016 for concentrations of several different heavy metals observed in the spill. None of these metals were present in high enough concentrations to raise concerns about fish consumption by humans, and the levels of all the metals tested by the Colorado Department of Public Health and the Environment probably represent background levels (CDPHE 2018). Similarly, there were no significant impacts on the abundance or community composition of benthic (bottom-dwelling) aquatic macroinvertebrates (mostly insect larvae and other invertebrate groups) (Roberts 2016).

The information from studies of floodplain vegetation, fish, and aquatic invertebrates corroborates the results of a three-year post-spill study released in August 2018 by San Juan Basin Public Health (SJBPH 2018). Based on extensive river water, well water, and sediment samples, the report concludes that the Gold King spill had no lasting impacts on water quality (SJBPH 2018).

SUMMARY

The broad view of the GKM spill presented here includes a deep time perspective beginning 1,800 million years ago. The GKM event and potential environmental impacts represent an interplay of the geologic history, ecology, and modern hydrologic systems. Geologic events created the foundation for the interaction of mining and drainage systems that define environmental conditions within the Animas River watershed today. Resultant AMD and ARD have led to the Animas River's dubious distinction as a high metal-load stream.

Ironically, the biggest impacts of the GKM spill may turn out to be mostly positive, especially since the short-term impact on the ecosystem was less than expected. Most important, it drew the public's attention to the long-term deterioration of water quality in the Animas River due to chronic heavy metal loading from AMD and ARD. The result is that now there is active monitoring of water quality (CDPHE 2018; EPA 2018; MSI 2018; SJBPH 2018) and aquatic life (MSI 2018; White 2016), and there is more information available to the community (ARCF 2018; CDPHE 2018; MSI 2018).

Unfortunately, though, heavy metal loading is only one of the challenges that limits the resilience of this shared resource. Lower stream flows due to lower precipitation (including snowpack), greater aridity, and greater human demand result in an increase in heavy metal concentrations as well as other pollutants. Greater aridity and higher temperatures are also associated with increased fire activity, as seen in the summers of 2002 and 2018. In both cases, ash from fires and muddy water from post-fire runoff caused fish die-offs (Romeo 2018; Simonovich 2018), which were not seen with the pulse of heavy metals from the GKM spill. This combination of stressors limits the resilience of our river systems, and their ability to provide vital ecosystem services to society is in jeopardy. Given this, it is imperative that we continue to work to address factors that degrade water quality and limit water quantity in the rivers and streams of the arid American West.

REFERENCES

Amlin, Nadine M., and Stewart B. Rood. 2002. "Comparative Tolerances of Riparian Willows and Cottonwoods to Water-Table Decline." *Wetlands* 22 (2): 338–346.

ARCF (Animas River Community Forum). 2018. *Animas River Community Forum*. https://www.animasrivercommunity.org/.

Blake, Johanna M. 2017. "Day 3B Road Log: Gold King Mine Release and Environmental Concerns." In *The Geology of the Ouray-Silverton Area*, edited by Karl E. Karlstrom, David A. Gonzales, Matthew J. Zimmerer, Matthew Heizler, and Dana S. Ulmer-Scholle, 59–65. New Mexico Geological Society 68th Annual Fall Field Conference Guidebook. Las Cruces: New Mexico Geological Society.

Burbank, Wilbur Swett, and Robert G. Luedke. 2008. *Geology and Ore Deposits of the Uncompahgre (Ouray) Mining District, Southwestern Colorado*. US Geological Survey Professional Paper 1753. Denver, CO: US Geological Survey.

Casadevall, Tom, and Hiroshi Ohmoto. 1977. "Sunnyside Mine, Eureka Mining District, San Juan County, Colorado: Geochemistry of Gold and Base Metal Ore Deposition in a Volcanic Environment." *Economic Geology* 72: 1285–1320.

Cather, Steven M. 2004. "Laramide Orogeny in Central and Northern New Mexico and Southern Colorado." In *The Geology of New Mexico: A Geologic History*, edited by Greg H. Mack and Katherine A. Giles, 203–248. Special Publication 11. Las Cruces: New Mexico Geological Society.

CDPHE (Colorado Department of Health and the Environment). 2018. *Animas River Reports: Data and Maps*. C. D. Environment, Producer. https://www.colorado.gov/pacific/cdphe/animas-river-water-quality-sampling-and-data.

Church, Stanley E., D. L. Fey, E. M. Brouwers, C. W. Holms, and Robert Blair. 2000. "Determination of Pre-Mining Geochemical Conditions, and Paleoecology in the Animas River Watershed, Colorado." Open-File Report 99-0038, United States Geological Survey 16. Denver, CO: US Geological Survey.

Church, Stanley E., J. Robert Owen, Paul von Guerard, Philip L. Verplanck, Briant A. Kimball, and Douglas B. Yager. 2007. *Understanding and Responding to Hazardous Substances at Mine Sites in the Western United States*, vol. 17: *The Effects of Acidic Mine Drainage from Historical Mines in the Animas River Watershed, San Juan County, Colorado: What Is Being Done and What Can Be Done to Improve Water* Quality, edited by Jerome V. DeGraff, 47–83. Geological Society of America Reviews in Engineering Geology. Boulder, CO: Geological Society of America. doi: https://doi.org/10.1130/REG17.

Clements, William H., Daren M. Carlisle, James M. Lazorchak, and Philip C. Johnson. 2000. "Heavy Metals Structure Benthic Communities in Colorado Mountain Streams." *Ecological Applications* 10 (2): 626–638.

Culver, Denise R., and Joanna M. Lemly. 2013. *Field Guide to Colorado's Wetland Plants*. Fort Collins: Colorado Natural Heritage Program.

Donahue, Maria Magdalena. 2016. "Episodic Uplift of the Rocky Mountains: Evidence from U-Pb Detrital Zircon Geochronology and Low-Temperature Thermochronology with a Chapter on Using Mobile Technology for Geoscience Education: Earth and Planetary Sciences." PhD dissertation, University of New Mexico, Albuquerque.

Donahue, Magdalena S., Karl E. Karlstrom, Mark Pecha, David A. Gonzales, and Rachel Price. 2012. "Insights into Paleogeography and Sedimentation during Early Stages of the Oligocene Ignimbrite Flare-up Using Detrital Zircon U-Pb Analyses on Sedimentary Inits in the Southern Rocky Mountains: Telluride Conglomerate and Blanco Basin Formation." *Geological Society of America Rocky Mountain Section, Abstracts with Programs* 44 (60): 20.

Dott, Cynthia, Gary L. Gianniny, M. J. Clutter, and Colin Aanes. 2016. "Temporal and Spatial Variation in Floodplain Aquifers and Riparian Vegetation on the Regulated Dolores River, Southwest Colorado, USA." *River Research and Applications* 32: 2056–2070. https://doi.org/10.1002/rra.3042.

Douhovnikoff, Vladimir, Joe R. McBride, and Richard S. Dodd. 2005. "Salix exigua Clonal Growth and Population Dynamics in Relation to Disturbance Regime Variation." *Ecology* 86 (2): 446–452.

EPA (Environmental Protection Agency). 2018. *Follow-Up Monitoring Data from Gold King Mine Incident*. https://www.epa.gov/goldkingmine/follow-monitoring-data-gold-king-mine-incident.

Fisher, Frederick. 1990. "Gold Deposits in the Sneffels-Telluride and Camp Bird Mining Districts, San Juan Mountains, Colorado." *US Geological Survey Bulletin* 1857-F: F12–F17.

Gillam, Mary L. 1998. "Late Cenozoic Geology and Soils of the Lower Animas River Valley, Colorado and New Mexico." PhD dissertation, University of Colorado, Boulder.

Gonzales, David A. 1997. "Crustal Evolution of the Needle Mountains Proterozoic Complex, Southwestern Colorado." PhD dissertation, University of Kansas, Lawrence.

Gonzales, David A., and William R. Van Schmus. 2007. "Proterozoic History and Crustal Evolution in Southwestern Colorado: Insight from U/Pb and Sm/Nd Data." *Precambrian Research* 154 (1–2): 31–70.

Gonzales, David A. 2015. "New U-Pb Zircon and 40Ar/39Ar Age Constraints on the Late Mesozoic to Cenozoic Plutonic Record in the Western San Juan Mountains." *Mountain Geologist* 52 (2): 5–41.

Gonzales, David A., and R. A. Larson. 2017. "An Overview of the Mineral Deposits of the Red Mountain Mining District, San Juan Mountains, Colorado." In *The Geology of the Ouray-Silverton Area*, edited by Karl E. Karlstrom, David A. Gonzales, Matthew J. Zimmerer, Matthew Heizler, and Dana S. Ulmer-Scholle, 133–140. New Mexico Geological Society 68th Annual Fall Field Conference Guidebook. Las Cruces: New Mexico Geological Society.

Gregory, Stanley V., Frederick J. Swanson, W. Arthur McKee, and Kenneth W. Cummins. 1991. "An Ecosystem Perspective of Riparian Zones." *Bioscience* 41 (8): 540–551.

Guido, Zackry S., Dylan J. Ward, and Robert S. Anderson. 2007. "Pacing the Post-Last Glacial Maximum Demise of the Animas Valley Glacier and the San Juan Mountain Ice Cap, Colorado." *Geology* 35 (8): 739–742. doi: https://doi.org/10.1130/G23596A.1.

Hereford, Richard, and Robert H. Webb. 1992. "Historic Variation of Warm-Season Rainfall, Southern Colorado Plateau, Southwestern U.S.A." *Climatic Change* 22: 239–256.

Hereford, Richard, Robert H. Webb, and Scott Graham. 2002. "Precipitation History of the Colorado Plateau Region, 1900–2000." US Geological Survey Fact Sheet. Denver, CO: US Geological Survey. http://geopubs.wr.usgs.gov/fact-sheet/fs119-02/.

Johnson, Brad, Mary L. Gillam, and Jared Beeton. 2017. "Glaciations of the San Juan Mountains: A Review of the Work since Atwood and Mather." In *Geology of the Ouray-Silverton Area*, edited by Karl E. Karlstrom, David A. Gonzales, Matthew J. Zimmerer, Matthew Heizler, and Dana S. Ulmer-Scholle, 195–204. New Mexico Geological Society 68th Annual Fall Field Conference Guidebook. Las Cruces: New Mexico Geological Society.

Johnson, Mark D., and Mary L. Gillam. 1995. "Composition and Construction of Late Pleistocene End Moraines, Durango, Colorado." *GSA Bulletin* 107 (10): 1241–1253. doi: https://doi.org/10.1130/0016-7606(1995)107<1241:CACOLP>2.3.CO;2.

Karlstrom, Karl E., David A. Gonzales, Matthew Heizler, and A. Zimmer. 2017. "40AR/39Ar Age Constraints on the Deposition and Metamorphism of the Uncompahgre Group, Southwestern Colorado." In *The Geology of the Ouray-Silverton Area*, edited by Karl E. Karlstrom, David A. Gonzales, Matthew J. Zimmerer, Matthew Heizler, and Dana S. Ulmer-Scholle, 83–90. New Mexico Geological Society 68th Annual Fall Field Conference Guidebook. Las Cruces: New Mexico Geological Society.

Kittel, Gwen. 2016. *A Vegetative Key to the Willows of Colorado*. Colorado Native Plant Society, March 6. https://conps.org/wp-content/uploads/2016/04/Kittel_Salix_Key_ver_3-2016-1.pdf.

Kluth, Charles F., and Peter J. Coney. 1981. "Plate Tectonics of the Ancestral Rocky Mountains." *Geology* 9: 10–15.

Leary, Ryan F., Paul Umhoefer, M. Elliot Smith, and Nancy Riggs. 2017. "A Three-Sided Orogen: A New Tectonic Model for Ancestral Rocky Mountain Uplift and Basin Development." *Geology* 45 (8): 735–738. doi: 10.1130/G39041.1.

Lipman, Peter W. 2006. "Geologic Map of the Central San Juan Caldera Cluster, Southwestern Colorado." Geologic Investigations Series I-2799, pamphlet 34. Denver, CO: US Geological Survey. https://doi.org/10.3133/i2799.

Lipman, Peter W. 2007. "Incremental Assembly and Prolonged Consolidation of Cordilleran Magma Chambers: Evidence from the Southern Rocky Mountain Volcanic Field." *Geosphere* 3: 42–70.

Lipman, Peter W., and Oliver Bachmann. 2015. "Ignimbrites to Batholiths: Integrating Perspectives from Geological, Geophysical, and Geochronological Data." *Geosphere* 11 (3): 705–743. doi:10.1130/GES01091.1.

Lipman, Peter W., Michael A. Dungan, L. L. Brown, and Allen Deino. 1996. "Recurrent Eruption and Subsidence at the Platoro Caldera Complex, Southeastern San Juan Volcanic Field, Colorado: New Tales from Old Tuffs." *GSA Bulletin* 108 (8): 1039–1055. doi: https://doi.org/10.1130/0016-7606(1996)108<1039:REASAT>2.3.CO;2.

Lipman, Peter W., Frederick S. Fisher, Harold H. Mehnert, Charles W. Naeser, Robert G. Luedke, and Thomas A. Steven. 1976. "Multiple Ages of Mid-Tertiary Mineralization and Alteration in the Western San Juan Mountains, Colorado." *Economic Geology* 71: 571–588.

Lipman, Peter W., and William C. McIntosh. 2008. "Eruptive and Noneruptive Calderas, Northeastern San Juan Mountains, Colorado: Where Did the Ignimbrites Come From?" *Geological Society of America Bulletin* 120 (7–8): 771–795. https://doi.org/10.1130/B26330.1.

Lipman, Peter W., Thomas A. Steven, Robert G. Luedke, and Wilbur S. Burbank. 1973. "Revised Volcanic History of the San Juan, Uncompahgre, Silverton, and

Lake City Calderas in the Western San Juan Mountains, Colorado." *US Geological Survey Journal of Research* 1: 627–642.

Lipman, Peter W., Matthew J. Zimmerer, and William C. McIntosh. 2015. "An Ignimbrite Caldera from the Bottom Up: Exhumed Floor and Fill of the Resurgent Bonanza Caldera, Southern Rocky Mountain Volcanic Field." *Geosphere* 11 (6): 1902–1947. doi: doi.org/10.1130/GES01184.1.

Lucas, Spencer G. 2017. "Triassic-Jurassic Stratigraphy in Southwestern Colorado." In *The Geology of the Ouray-Silverton Area*, edited by Karl E. Karlstrom, David A. Gonzales, Matthew J. Zimmerer, Matthew Heizler, and Dana S. Ulmer-Scholle, 83–90. New Mexico Geological Society 68th Annual Fall Field Conference Guidebook. Las Cruces: New Mexico Geological Society.

Luedke, Robert G., and Wilbur S. Burbank. 1969. "Geology and Ore Deposits of the Eureka and Adjoining Districts, San Juan Mountains, Colorado." US Geological Survey Professional Paper 535. Denver, CO: US Geological Survey. https://doi.org/10.3133/pp535.

Luedke, Robert L., and Wilbur S. Burbank. 2000. "Geologic Map of the Silverton and Howardsville Quadrangles, Southwestern Colorado." US Geological Survey Map I-2681. Denver, CO: US Geological Survey.

Magliozzi, Chiara, Robert Grabowski, Aaron I. Packman, and Stefan Krause. 2017. "Scaling down Hyporheic Exchange Flows: From Catchments to Reaches." *Hydrology and Earth System Sciences Discussions*. doi: doi:10.5194/hess-2016-683.

Mahoney, John M., and Stewart B. Rood. 1998. "Streamflow Requirements for Cottonwood Seedling Recruitment: An Integrative Model." *Wetlands* 18 (4): 634–645.

Maltby, Edward, and Mike C. Acreman. 2011. "Ecosystem Services of Wetlands: Pathfinder for a New Paradigm." *Hydrological Sciences Journal* 56 (8): 1341–1359.

Marzonie, Magena. 2017. "Ecological and Chemical Analysis of Heavy Metal Transduction in Salix exigua on the Animas and Florida Rivers." *Metamorphosis*. https://metamorphosis.coplac.org/index.php/metamorphosis/article/view/44.

McKnight, Edwin T. 1974. *Geology and Ore Deposits of the Rico District, Colorado*. US Geological Survey Professional Paper 72. Denver, CO: US Geological Survey.

MSI (Mountain Studies Institute). 2018. *Animas River Monitoring*. http://www.mountainstudies.org/animasriver/.

Naiman, Robert J., Henri Decamps, and Michael Pollock. 1993. "The Role of Riparian Corridors in Maintaining Regional Biodiversity." *Ecological Applications* 3 (2): 209–212. doi:https://doi.org/10.2307/1941822.

Nilsson, Christer, and Kajsa Berggren. 2000. "Alterations of Riparian Ecosystems Caused by River Regulation." *Bioscience* 50 (9): 783–792.

Pecha, Mark E., George E. Gehrels, Karl E. Karlstrom, William R. Dickinson, Magdalena S. Donahue, David A. Gonzales, and Michael D. Blum. 2018. "Provenance of Cretaceous through Eocene Strata of the Four Corners Region: Insights from Detrital Zircons in the San Juan Basin, New Mexico and Colorado." *Geosphere* 14 (2): 785–811. doi: https://doi.org/10.1130/GES01485.1.

Rivera, Tiffany A., Mark D. Schmitz, James L. Crowley, and Michael Storey. 2014. "Rapid Magma Evolution Constrained by Zirxon Petrochronology and 40AR/39Ar Sanidine Ages for the Huckleberry Ridge Tuff, Yellowstone, USA." *Geology* 42 (8): 643–646.
Robb, Laurence. 2005. *Introduction to Ore-Forming Processes*. Malden, MA: Blackwell.
Roberts, Scott. 2016. *Animas River 2015 Benthic Macroinvertebrate (BMI) Report—Gold King Mine Release Monitoring*. Mountain Studies Institute—Animas River Monitoring. https://static1.squarespace.com/static/53bc5871e4b095b6a42949b4/t/5a0caf2ef9619a2a369292d4/1510780794119/EPA_BMI_Report_20170309.pdf.
Romeo, Jonathan. 2018. "Resilient River: Battered in Recent Years by Ash and Mine Waste, the Animas Has Bounced Back Before." *Durango Herald*, August 4. https://durangoherald.com/articles/235122.
Rood, Stewart B., Sarah G. Bigelow, and Alexis A. Hall. 2011. "Root Architecture of Riparian Trees: River Cut-Banks Provide Natural Hydraulic Excavation, Revealing That Cottonwoods Are Facultative Phreatophytes." *Trees* 25: 907–917.
Sfakianakis, Dimitris G., Elisavet Renieri, M. Kntouri, and Aristides M. Tsatsakis. 2015. "Effect of Heavy Metals on Fish Larvae Deformities: A Review." *Environmental Research* 137: 246–255.
Simonovich, Ryan. 2018. "CPW: Thousands of Fish May Be Dying from 416 Fire Ash in Animas River." *Durango Herald*, July 17. https://durangoherald.com/articles/232233.
SJBPH (San Juan Basin Public Health). 2018. *Water Quality*. S. J. Health, Producer. http://sjbpublichealth.org/waterquality/.
Smith, Walter. 1986. *The Effects of Eastern North Pacific Tropical Cyclones on the Southwestern United States*. National Oceanic and Atmospheric Administration, US Department of Commerce. Salt Lake City, UT: National Weather Service.
Somers, Preston, and Lisa Floyd-Hanna. 1996. "Wetlands, Riparian Habitats, and Rivers." In *The Western San Juan Mountains: Their Geology, Ecology, and Human History*, edited by R. Blair, 175–189. Boulder: University Press of Colorado.
Spencer, Albert W., and William H. Romme. 1996. "Ecological Patterns." In *The Western San Juan Mountains: Their Geology, Ecology, and Human History*, edited by Rob Blair, 129-142. Boulder: University Press of Colorado.
Steven, Thomas A., and Peter W. Lipman. 1976. "Calderas of the San Juan Volcanic Field, Southwestern Colorado." US Geological Survey Professional Paper 958. Denver, CO: US Geological Survey. https://doi.org/10.3133/pp958.
Timmons, Stacy, Ethan Mamer, and Cathryn Pokorny. 2016. *Groundwater Monitoring along the Animas River, New Mexico: Summary of Groundwater Hydraulics and Chemistry from August 2015 to June 2016*. Aquifer Mapping Program, New Mexico Bureau of Geology and Mineral Resources. https://geoinfo.nmt.edu/resources/water/amp/brochures/FTR_Animas_River_Sept_2016_LR.pdf.

USGS (US Geological Survey). 1912–present. *USGS 09361500 Animas River at Durango, CO.* US Geological Survey, Producer. National Water Information System: Real Time Water Data. https://waterdata.usgs.gov/co/nwis/uv?site_no=09361500.

Ward, James V., Klement Tockner, David B. Arscott, and Cecile Claret. 2002. "Riverine Landscape Diversity." *Freshwater Biology* 47: 517–539.

Webb, Robert H., Stanley A. Leake, and Raymond M. Turner. 2007. *The Ribbon of Green.* Tucson: University of Arizona Press.

White, James N. 2016. *Conference on Environmental Conditions of the Animas and San Juan Watersheds.* https://animas.nmwrri.nmsu.edu/wp-content/uploads/SpeakerPresentations/JimNWhite.pdf.

Williams, Christopher A., and David J. Cooper. 2005. "Mechanisms of Riparian Cottonwood Decline along Regulated Rivers." *Ecosystems* 8: 1–14.

Winter, Thomas C., Judson W. Harvey, O. Lehn Franke, and William M. Alley. 1998. "Ground Water and Surface Water: A Single Resource." US Geological Survey Circular 1139. Denver, CO: US Geological Survey.

WWA (Western Water Assessment). 2015. *Intermountain West Climate Dashboard.* http://wwa.colorado.edu/climate/dashboard.html.

Yager, Douglas, and Dana J. Bove. 2002. "Generalized Geologic Map of the Upper Animas River Watershed and Vicinity, Silverton, Colorado." US Geological Survey Miscellaneous Field Studies Map 2377. Denver, CO: US Geological Survey. https://doi.org/10.3133/mf2377.

5

Watershed Consciousness

The Animas River and a Sense of Place

PETE MCCORMICK

In the spring of 2016, I presented an earlier version of this chapter at the Annual Meeting of the American Association of Geographers in San Francisco (McCormick 2016). I am a cultural geographer by training, and the audience was made up of other geographers—members of a discipline that has been defined in part for decades by a broad interest in places, landscapes, and human interaction with the natural world. My talk was a broad overview of the impact of the Gold King Mine (GKM) spill on the sense of place of the Durango, Colorado, region. My initial thesis was that the spill, which dominated the headlines of the local newspaper—the *Durango Herald*—and captured the local, national, and international media's attention, held similar company with other events and landscape features that centrally located the river within the town's collective identity and imagination. In this follow-up chapter, I ask some fairly common questions a geographer might ask: How is this integral part of the southwestern Colorado landscape understood by the region's inhabitants, and how does the mine spill 50 miles upstream near Silverton relate to it?

DOI: 10.5876/9781646421756.c005

In 2016, I argued that this event—regardless of its short- or long-term environmental effects on the region—was one of several events and processes that have placed the river at the front of the consciousness of the communities that live within the watershed's boundaries.[1] My initial hunch was that the 2015 spill had similar company in the development of the region's sense of place—these other events include the Animas River flood of 1911, the dismantling of the Uranium Smelter and subsequent remediation in the late twentieth century, and the building of the US Bureau of Reclamation's final crown jewel: the Animas–La Plata Project (ALP). These landscape symbols triggered tremendous community reaction and were driven by long-standing legacies of natural resource–based economics that are deeply entrenched in the watershed's economy, laws, legal codes, and, frankly, its historical and contemporary self. When placed in historical and geographical context, it is easier to understand why the 2015 spill elicited a strong response. It is much easier as well to see how the river itself has shaped the region's identity.

A WATERSHED AND SENSE OF PLACE

For geographers, the title of this chapter may bring to mind James Parsons (1985) and Stephen Frenkel (1994) and their discussions of bioregionalism, environmental determinism, and "watershed consciousness" in the 1980s and 1990s. While the Gold King spill did not result in a call for local residents to re-inhabit the Animas Basin in a more "sustainable and appropriate" manner—as early advocates of bioregionalism, for example, may very well have done—it did evoke such an overwhelming reaction from the local media and community members that it will not quickly be forgotten. As previous chapters in this volume have noted, the spill is entangled in incredible policy complexity, jockeying by stakeholders, and political maneuvers by outsiders; further, it has caused environmental, economic, and social harm. It interrupted the natural rhythms and cycles of life along and in the river. Just as enthralling (for the geographer) is the role the event played in solidifying the river in the real and imagined landscape of the basin. The spill put the Animas River on the front page and in the spotlight for residents and outsiders alike. While the spill vanished quickly from the headlines in national and international media, the river remains on the minds of many of the people who inhabit its basin.

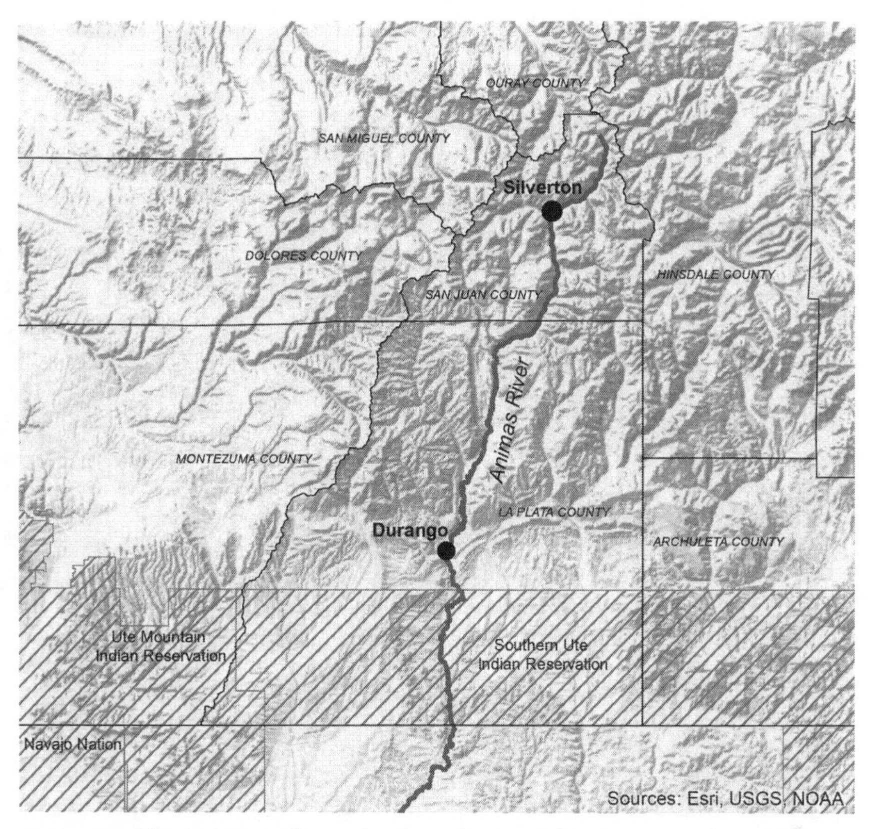

FIGURE 5.1. The Animas Valley and surrounding communities in southwestern Colorado. *Courtesy*, Jesse W. Tune, Department of Anthropology, Fort Lewis College, Durango, Colorado.

Durango, which means "water town" or "water place" in Basque,[2] sits in the middle of the Animas Basin halfway between the headwaters above Silverton, Colorado, and the junction of the Animas with the larger San Juan River to the south in New Mexico (figure 5.1). It sits just below moraines deposited by glaciers that carved the upper Animas Valley and at the transition zone (ecotone) between the foothills and the high desert of the Colorado Plateau (Blair 1996). Until the early 1980s, the Animas Basin was dominated by mining, ranching, and forestry; since then, it has become a destination landscape that is home to nearly 60,000 retirees, telecommuters, ski bums, mountain bikers, college students, kayakers, and climbers (Smith

FIGURE 5.2. The town of Durango with the Animas River in the foreground ca. 1915. *Courtesy*, Center of Southwest Studies, Fort Lewis College, Durango, Colorado. I am deeply indebted to Nic Kendziorski, the archives manager, for help with this and the next five figures.

1980). The Animas is an iconic part of southwestern Colorado, rivaled only by the 13,000 foot peaks of the La Plata Range of the San Juan Mountains. From a geographic and marketing perspective, this is a special place, a quintessential southwestern landscape (figure 5.2). Yet local understanding of the landscape is much more than a backdrop, a landscape painting, or an online ad. In addition to the GKM spill, the three other events involving the river (the Uranium Smelter, the flood of 1911, the Animas–La Plata Project) placed the river at the center of the community's sense of place. These are four events with significant health, policy, and historical ramifications, but they all center around a geographic location (Durango) and a watershed (the Animas Basin) and ultimately are place-making events and icons. Place was and remains very much at the center of many environmental issues related to the GKM spill. Simply put, the Animas River is that place.

Geographer Yi Fu Tuan (1974) outlined metaphors and contexts for understanding place—and the Animas River—in the cultural context. As he wrote in a later publication, "In ordinary usage, place means primarily two things:

one's position in society and one's spatial location" (Tuan 1996, 445). For Tuan, place had a much deeper meaning and role in the human experience, and his most influential work highlighted human attachment and response to place. This early theoretical framework for a humanistic understanding of geography beyond region, landform, and location acknowledged that places have their own personalities and also elicit emotional change. Tuan demarcated places as having a *spirit* and a *sense*. Spirit of place evokes personality and greater emotional charge than sheer geographic location and suggests that places have their own personality. Sense of place, in contrast, is an understanding of place beyond the "spirit" of the aesthetics and suggests that one has a much deeper understanding of place—including familiar sounds, colors of light, and the texture of pavement, for example—because of a much longer period of contact (Tuan 1974). For the tourist and the first-time resident, the Animas functions largely as part of the larger aesthetic and marketing draw of the greater Southwest and the Four Corners region. Perhaps the only other landscape symbols that rival the river in this region are the San Juan Mountains. The two combined contribute to the region's spirit of place. However, just like the mountain slopes its tributaries drain, the Animas is also a very important public symbol, and that public symbol is directly linked to residents' sense of place.[3]

THE ANIMAS AS PUBLIC SYMBOL

The Animas, both literally and figuratively, defines a tremendous amount of the region's physical and cultural landscape. The river and its tributaries are companions with the surrounding mountains on scales of geographic grandeur and historical, environmental, economic, and cultural importance. All of the geographic features of this valley have some cultural significance, but the Animas has a special place in the subjective landscape. The river slithers and bounds down from its headwaters above 14,000 feet and meanders across a broad glacially carved valley en route to Durango. The river is also a major player in the local business scene, in public spaces, and on street signs. It is a public symbol, as defined by Tuan and other geographers, because it has high visibility on the landscape—it literally dissects both La Plata County and the City of Durango in north-to-south fashion and is bounded on all sides by parklands, bridges, and playgrounds. The public eye was not always on Durango's Animas because of its recreational prowess, ecological diversity, and serene shores. It was made more visible and important through controversy and the investment of human emotion, time, and, as Tuan would note, spirit.

Understanding the importance of the river to the physical landscape of Durango and the surrounding area goes beyond simple cartography. A geographer, planner, geologist, or environmental consultant may point out that the contours of the river's floodplains determine much of the layout of the grids of the valley's original town sites: Animas City and Durango. Both sit at roughly 30 degree angles slanted northeast paralleling the flow of the Animas (figure 5.3). The route of the Durango & Silverton Narrow Gauge Railway—a historic narrow gauge tourist train that draws thousands of visitors annually—quite naturally follows the contours of the river on its way to Silverton.

The Animas is also located in several other places within the watershed. Dotted across the landscape of the Durango area are dozens of locations whose names feature some form of Animas and River (or Rio). Twelve streets in the valley are named explicitly (or in some likeness of) Animas, River, or Rio, including Animas River Drive, Animosa Lane, and the main corridor through downtown and southern Durango, the Camino del Rio (the River Way). Seventy businesses and public offices with the

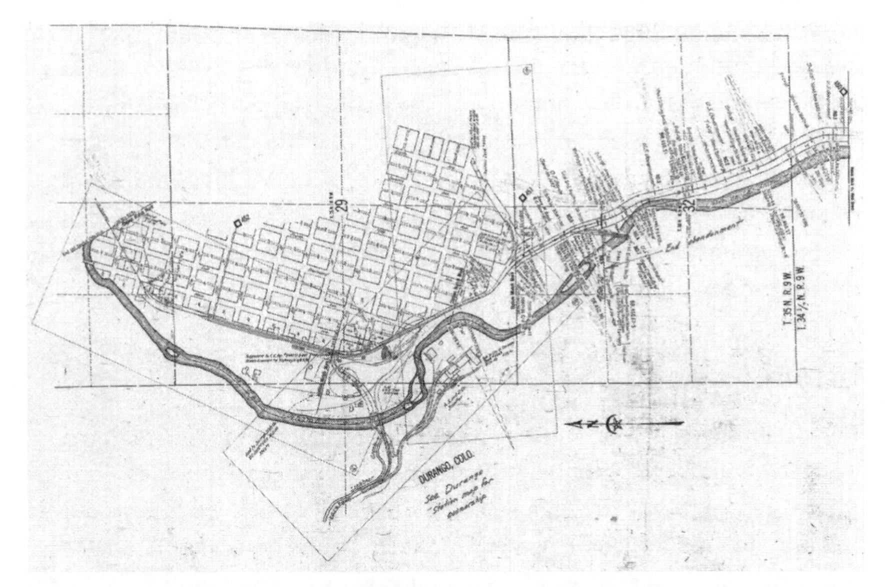

FIGURE 5.3. Denver & Rio Grande Railroad map of the railroad, the river, the smelter, and the original town plat of Durango ca. 1915. *Courtesy*, Center of Southwest Studies, Fort Lewis College, Durango, Colorado.

name Animas in their title are listed in the local telephone directory, ranging from animal shelters to liquor stores to medical groups: the Animas Shelter–Humane Society, Animas Wine and Spirits, and the Animas Surgical Hospital. There is also an elementary school, Riverview; and a charter high school, Animas High. One of the largest live music venues, which also plays independent films, is in the heart of downtown Durango: the Animas City Theatre. The Animas is literally everywhere. River is too, but it is not nearly as common in local business directories, with eighteen listings. As a point of reference, the terms *Mountain* and *La Plata* combine for a similar number of business and public titles: eighty-five compared to Animas and River's total of eighty-eight.[4] For locals, the frequent use of the term *Animas* is so ordinary that to suggest it has greater meaning would be nonsensical. However, to push the argument a bit further, the largest paved trail in the city's extensive hiking, biking, and running trail system—the Animas River Trail—straddles the Animas for nearly 8 miles from north to south. It cuts through the heart of the city and is a project that has been forty years in the making. It runs through ten city parks and required the building of numerous bridges, underpasses, and a tunnel (City of Durango 2017; Turner 2017). In addition, one of the many festivals in town is Animas River Days, which celebrates the city's hydrologic heritage through a series of events including numerous kayaking competitions, raft parades, a river rodeo, a dance, and a film night. The Animas also offers gold medal fishing waters; and numerous rafting, kayaking, and fishing outfitters and supply stores line downtown streets.

FIELDS OF CARE

The Animas River is more than a hydrological system, the center of a watershed, and a recreational juggernaut. This public symbol also exudes what Tuan (1996, 451) recognized as a "field of care," which he notes is "also a place if the people are emotionally bound to their material environment and, further, they are conscious of its identity and spatial limit." For some reason, the Animas elicits deep emotions and has its own personality, prescribed to it by individual members of the local community but given greater context and meaning by relationships the community has developed with it. Several nonprofit, grassroots organizations advocate for environmental responsibility and awareness in the region; many of them focus much of their attention on the Animas. This indicates public awareness and action. An event in the fall of 2012 is an illustration. Over 9,000 local residents formed a continuous human chain along the Animas River Trail from the city's north terminus to the south to celebrate the completion of the trail. During the forty-year span of its development, a tremendous amount of planning, collaboration, financing, and persistence was required. The celebration event was equally involved, as it required the work of law enforcement, emergency services, service clubs, churches, sports teams, and local businesses. The event was intended to officially recognize the completion of the trail; but, as the organizer said, "it brought us together literally and symbolically" and was a way to reconnect community members even in times of difference (Turner 2017). The event was intended to connect community members; it also reconnected them to an important field of care and a public symbol in which they have invested a tremendous amount of cultural capital.

BACKSTORY

People in the Animas Basin should be used to environmental disasters. After all, Durango and southwest Colorado have witnessed several mining-related disasters and massive wildfires during the past century. It was not that different 100 years ago, either. I have used the contemporary landscape (both symbolic and real) as evidence of cultural significance. Historical geography can provide the context: newspapers, archives, and secondary sources, for example, tell us a *backstory* that rests in collective memory. There are clues here as to why Animas is adored by its neighbors. While they are the subjects of dozens of headlines of the past, the ALP, the flood, and the smelter haven't been forgotten. Not unlike the Animas, they are also historic public symbols of the town's relationship with the river.[5]

In 1978, the Uranium Mill Tailings Radiation Control Act was passed, and the US Environmental Protection Agency and the US Department of Energy were charged with cleaning up the most dangerous of the former locations that processed uranium for the Manhattan Project and the Cold War. This included the former smelter site along the Animas River just south of Durango's central business district (figure 5.4), which was finally cleaned up between 1986 and 1991 after several years of discussions and failed attempts at mitigating at the state, national, and local levels. Local sentiment on removing the last vestige of an economic mainstay (smelting) finally shifted toward cleanup after residents realized that the tailings could be detrimental to their health and were, in fact, radioactive (Center of Southwest Studies n.d.; US Department of Energy n.d.) (figure 5.5).[6] The local Chamber of Commerce was vital in getting the remediation process going, as the site and its iconic smokestack were viewed as blight by tourists. The uranium and vanadium smelters were mainstays to the local economy and ways of life. They employed hundreds of residents, and during the mid-twentieth century, students skied down the tailings piles. Durango's air was often full of tailings—yellow cake—during gusty spring days. So central were the smelter and its tailings to the town that a lot of the tailings ended up in the concrete and asphalt of streets or in foundations and basements of businesses and homes over the course of the early twentieth century (Center of Southwest Studies n.d.; Smith 1980; Thompson 2018). The legacy of smelting—which included iron, vanadium, and uranium—came to a close with the mediation of the location in the early 1990s, but 100 years earlier it had already left its

FIGURE 5.4. Durango Uranium Smelter on a postcard ca. 1915. *Courtesy*, Center of Southwest Studies, Fort Lewis College, Durango, Colorado.

mark on the history, economy, environment of the Animas Basin. The coal mines directly surrounding the junction of Lightner Creek and the Animas, as well as the mines near the headwaters near Silverton, led to the moniker "Smelter City" by the 1890s (Rohrabacher 1901; Smith 1980); and the valley had already become toxic. Historian Duane Smith (1980, 93) notes: "The Animas River remained polluted, despite fines and surveillance, and smelter smoke, intensified by coal and wood smoke, sullied Durango's deep blue sky and sharpened its distinctive aroma."

Durango at the time had a reputation as being a "hard place," and while it did not set a precedent for future environmental policies and discussions about segregation and social inequality, it certainly could have. The legacy of mining and smelting had already created major environmental problems at the confluence of the Animas and Lightner Creek. Those tailings piles, the toxic water, and air from the smokestacks also hovered over makeshift housing in labor camps including Mexican Flats and Chihuahua, in the 100-year floodplain, and in the Southside just above. These neighborhoods were occupied by Mexican, Hispano,[7] and Eastern European laborers who had traditionally worked at the smelters, in the mines, in ranching, and in the agricultural fields (Smith 1980). They were in the direct line of fire from the

FIGURE 5.5. The Durango Uranium Smelter at the base of Smelter Mountain ca. 1915. The Animas River is in the foreground. *Courtesy*, Center of Southwest Studies, Fort Lewis College, Durango, Colorado.

smoke and airborne contaminants from the tailings piles, and in October 1911, they were hit by the largest flood in the region's historical record. The river flow rose to 25,000 cubic feet per second—usual highs during October are just over 200 cubic feet per second—after 3 to 4 inches of rain fell in the basin from Silverton south to Durango (figure 5.6). The floodwaters inundated the community, bringing sediment and tailings and toxins from the dozens of mines in the headwaters and depositing them throughout the upper valley and the town (Butler 2011; Smith 1980; Thompson 2018). Four feet of water covered much of the lower part of the original town site. Closer to the smelter, the Mexican, Hispano, and Eastern European communities were partially, if not totally, submerged. Once the water receded and the debris was removed, unregulated development soon followed up and down the river. It was not until the 1970s that the city and county developed comprehensive

FIGURE 5.6. The Ninth Street Bridge during the 1911 flood. *Courtesy*, Center of Southwest Studies, Fort Lewis College, Durango, Colorado.

land management codes and issues of floodplain development and Federal Emergency Management Agency (FEMA) regulations were adopted and enforced.[8] Little was done to address the segregation of the working classes: Mexican Flats was removed and became a park in the 1970s, Chihuahua transitioned into a mobile home community, and the Southside was gentrified in the early 2000s. This added to the growing legacy of environmental injustice in the region, and little was said about the sediment and contaminated water that would flow downstream through the heart of the Southern Ute Reservation and into the Navajo Nation.

The ALP, which was authorized in 1968 through the Colorado River Basin Project Act, became the center of public debate by the mid-1970s when it, along with several others, were put on President Jimmy Carter's hit list of Western reclamation projects. The project, in short detail, was to divert a miniscule amount of water from the Animas River and pump it uphill into the Ridges Basin, where it would be stored in an earthen-dam reservoir for agricultural, industrial, and residential use (Smith 1980; Thompson 2018, 97).

Led by the belief in the false promise that water was the cure-all for the arid and semiarid West, proponents believed agriculture and urban development would be secured. The fight over the ALP was raucous, and the community was divided. Several spirited community meetings and the saga itself captured *Herald* headlines and piqued the interest of a filmmaker. The rich plot of the documentary *Cowboys, Indians, and Lawyers* laid out the twisted tale of conflict over development, mining, and agriculture (Dengel and Oppenheimer 2006). The water was to be (and now is) pumped from the Animas at the site of the original smelter and then pushed uphill into Ridges Basin behind (the aptly named) Smelter Mountain, where 125,000 acre-feet of water fill Lake Nighthorse—which, before being flooded, was home to a working ranch, had dozens of ancestral Puebloan burials, and was part of a wildlife refuge (Ellison 2009; Potter 2006; Thompson 2018). All of this lays directly below (downstream) the area where the tailings from the uranium cleanup were deposited (on top of Smelter Mountain). The ALP did not open until 2011, and it was not until the Ute Water Settlement of 1988 and a subsequent amendment in 2000 that the issue was resolved. Water was first pumped uphill to the reservoir in 2011 (US Bureau of Reclamation n.d.). In August 2015, the pumping plant was shut down because a large orange plume of water was headed its way from the north.

FROM BACKSTORY TO HEADLINE

The GKM spill headlined the front page of the *Durango Herald* on August 6, 2015: CATASTROPHE! In the days after, catastrophe was followed by a series of often repeated caricatures of the Animas River, from disaster to dangerous to concerns about environmental and human health (Olivarius-McAllister 2015). A day earlier, the Environmental Protection Agency reported that 1 million gallons of toxic water had spilled from the GKM and would leave Cement Creek, a tributary of the Animas, and head south toward Durango. A few days later, the size of the spill tripled to 3 million gallons. Aluminum, antimony, silver, arsenic, iron, cobalt, chromium, barium, and cadmium in an ochre-yellow plume entered the river. Local authorities, research institutes, and the media scrambled; so did geologists, biologists, and geographers. Television crews from Albuquerque and Denver stations and affiliates of international networks including Al

Jazeera, CNN, and NBC were soon on the scene. What unfolded was a sensationalized media spectacle in which the technological disaster unfolded as an environmental and economic nightmare for a region usually represented to the outside world as an environmental, cultural, and historical playground. Simply put, a major landscape *backstory*—the legacy of mining in the Animas River Basin and the political, economic, and perceptual drivers of its importance to the region's sense of identity—became a front-page headline (for the impact of media coverage on public policy, please see chapter 3, this volume).[9]

Despite the fact that natural disasters are fairly common (or at least significant and of high importance) in Durango, the local on-the ground reaction, as witnessed in the news and in conversation, appeared to be chaotic at best. It did not appear to follow any established regional or national protocol for hazard management or mitigation. Conflicting reports on the size and magnitude of the spill and the levels of short-term and long-term toxicity obfuscated public understanding of the issue. The sheriff of La Plata County closed the river to all access. Water was shut off for all communities along the river, including Durango and Fort Lewis College. Ditch managers, ranchers, and farmers in the valley north of Durango scrambled to shut head gates to several ditches north of town. South of Durango, the Southern Ute Tribe and the Navajo Nation declared states of emergency and threatened multi-million-dollar lawsuits. Within twenty-four hours, local scientists began testing waters before the spill arrived and continued once the orange plume was within city limits. For nearly a week, the Animas Basin appeared to be in chaos.

After a week, outside news reports and their news crews disappeared from the scene. Updates on the quality of the water (and its color) were somewhat common in regional newscasts and in Western papers, including the *Denver Post*, the *Albuquerque Journal*, and the *High Country News*. However, the spill dominated local headlines for months to come. In addition to the numerous caricatures of the spill early on, local newspapers continued to look at long-lasting environmental, social, and emotional effects. There were also stories about community-wide meetings, special county commission meetings, visits from public officials, planning sessions, and information sessions (Marcus 2015). There were also several healing ceremonies, including one conducted by the Southern Ute Tribe (Shinn 2015).

I started tracking local newspaper coverage of the spill for a talk I gave at regional geography meetings in San Antonio in the fall of 2015 (McCormick 2015). It was a very rich data set and was abundant and consistent for months after the event—reflecting public interest and, ultimately, concern. Coverage continued three years later. The *Durango Herald* has included the GKM spill in no fewer than 543 articles since August 15, 2015. The readership of the *Herald* has been provided at least every other day with information on the spill and the effects it has had on environmental quality and health, on public policy, and on the opinions of readers from the valley and beyond. It is very safe to say that the local readership is well-versed on the spill and what it did to the community; and many of those readers have deep convictions and strong opinions. And rightly so because the Animas is a primary point of reference, and it was being disrupted. Simply put, the GKM spill—like the others—put the river, its environment, and mining front and center on computer screens, in newspapers, in casual conversation, and in public debate. An analysis of the articles in the *Herald* the first five months after the spill indicated that the three words most often used regarding the spill were change, water, and Animas. Readers were being told that the water and the Animas were changing (McCormick 2015).[10] The possibility of the latter appeals to residents' senses because the Animas is something they identify with and know well. In the same time period in which over 500 articles on the GKM were printed (since August 2015), over 1,200 articles in the *Herald* alone mentioned the word *Animas*. When combined with landscape morphology, place names, and business data, these media data suggest to me that high visibility, heightened awareness, and concern have contributed to the river's meaning and the depth of local care for it (Tuan 1974).

A SENSE OF PLACE

Geographers who have dealt with landscapes of natural resource extraction, particularly mining and its associated industries, have long decried them as hard, derelict, isolated, and raw (Francaviglia 1991; Robertson 2006). David Robertson (2006) argues that these kinds of landscapes have their own particular meaning for their residents, and very often these landscapes are central—regardless of their ugliness to the outside observer—to the identities of the communities that inhabit them. Durango has done an amazing job

FIGURE 5.7. The Animas River in southern Durango looking north toward the San Juan Mountains. *Courtesy*, Center of Southwest Studies, Fort Lewis College, Durango, Colorado.

of brushing its hard, derelict legacy under the carpet in terms of aesthetics; much of it has literally been torn down and removed. Yet the legacy of natural resource–based economies and disasters in the Animas Basin (and the rest of the West for that matter) has required that communities respond with disaster management, new policy, and legal code. The GKM spill resurrected the ghosts of Durango's geographical past and parts of its personality that we don't often see. If residents didn't witness the disaster firsthand, it was available in every newsstand and on every coffee table and computer screen (figure 5.7).

Geologists and climatologists tell us there is more of this to come. For example, in the summer of 2018, the 416 Fire exploded just north of Durango on the west bank of the Animas River and burned over 54,000 acres of national forest, private land, and wilderness. It caused smoke to fill the Animas Valley. This had happened before, in 2002, when the Missionary Ridge Fire engulfed over 100,000 acres on the east flank of the river. Both fires caused tremendous ecological damage. Both caused massive flash floods and debris flows that destroyed public property, and the latest fire killed thousands of fish in the Animas (Shinn 2018). In 2018, the debris flows turned the river black and brown. There are hundreds of abandoned mines in the upper Animas Basin, which gives residents pause and concern over the quality of water in the river and the likelihood of the next spill.

A lot has been said about the spill in this volume, in newspapers, and in thousands of local conversations. The discussions will likely never end, and Durango and the Animas will always be involved. The legacy of mineral extraction and the vulnerability of Western mining landscapes to human and natural disasters will never go away, and neither will the threats of drought and fire. In the wake of the spill, we may see new precedents set on how communities deal with such disasters in terms of policy and public infrastructure. What we do know is that those events will likely tell us a lot about a place and its personality.

NOTES

1. Jonathan Thompson (2018) makes a similar argument in his book on the Animas River. We clearly were thinking about this place-based stuff at the same time. His very insightful, enjoyable book came out before we went to press with this chapter, and I have referenced his book numerous times where his ideas intersect with mine—particularly in portions of my presentations at national and regional meetings in 2015 and 2016.

2. The City of Durango claims this translation on its government website: https://www.durangogov.org/Index.aspx?NID=274.

3. My methodological and theoretical framework is based on practice, training, and experimenting. My theoretical location is broadly humanistic and deeply informed by work by cultural geographers and those in the newly anointed "geohumanities." I use observation, experience, and analysis of local cultural artifacts (maps, phone books, newspapers, and the landscape itself) to construct this tale.

4. These data are primarily from *Directory Plus* (2018) and the *Durango Herald*.

5. My talks at national and regional geography meetings have looked at both media coverage (McCormick 2015) and the backstory of resource development and sense of place (McCormick 2016).

6. I rely heavily here on a senior thesis written under my supervision (Schnarch 2016).

7. The Hispano (or Indo-Hispano) population is the historic Spanish-speaking population of northern New Mexico and southern Colorado. Its roots are primarily Iberian in nature, including large numbers of families from Andalusia, many of whom, while now Catholic, converted from Sephardic Judaism to avoid the Inquisition.

These families moved into New Mexico in the seventeenth and eighteenth centuries and intermarried with indigenous communities (particularly Diné [Navajo] and Puebloan) and French traders and trappers (Hordes 2005; Nostrand 1992).

8. Both the City of Durango and La Plata County began some form of land-use regulation by the 1970s, including floodplain development restrictions required by FEMA.

9. This and the following paragraph are directly culled from *Durango Herald* coverage of the Gold King Mine spill. I have directly referenced articles in this section. There were hundreds of articles on the spill in the *Herald*, however, and it is impossible to reference each one individually. *Herald* coverage of the spill is located on the Gold King page of its website, https://durangoherald.com/tags/gold-king-mine. Also see chapter 3 in this volume, which outlines the media coverage and points to the effects of media on policy and public opinion.

10. The analysis using enVivo looked at five months of *Durango Herald* coverage.

REFERENCES

Blair, Robert. 1996. "Origin of Landscapes." In *The Western San Juan Mountains*, edited by Robert Blair, 3–17. Boulder: University Press of Colorado.

Butler, Anne. 2011. "Durango's Worst Flood Ever." *Durango Herald*, October 8. https://durangoherald.com/articles/9779.

Center of Southwest Studies. n.d. M 008 Durango, Colorado, Uranium Mill Tailings Collection. Center of Southwest Studies, Durango, CO.

City of Durango, Colorado. 2017. *Durango Animas River Corridor Management Plan*. Durango, CO: City of Durango.

Dengel, Julia, and Jonathan Oppenheimer. 2006. *Cowboys, Indians, and Lawyers*. Dir. Julia Dengel. Oley, PA: Bullfrog Films.

Directory Plus: Southwest Colorado. 2018. Durango, CO: Ballantine.

Ellison, Brian A. 2009. "Bureaucratic Politics, the Bureau of Reclamation, and the Animas–La Plata Project." *Natural Resources Journal* 49 (Spring): 366–371.

Francaviglia, Richard. 1991. *Hard Places: Reading the Landscape of America's Historic Mining Districts*. Iowa City: University of Iowa Press.

Frenkel, Stephen. 1994. "Old Theories in New Places: Environmental Determinism and Bioregionalism." *Professional Geographer* 46 (3): 289–295.

Hordes, Stanley. 2005. *To the End of the Earth: A History of the Crypto-Jews in New Mexico*. New York: Columbia University Press.

Marcus, Peter. 2015. "Hickenlooper Drinks Animas River Water to Make a Point." *Durango Herald*, August 12. https://durangoherald.com/articles/94034.

McCormick, Peter. 2015. "On the Brink of Disaster: Local Media and the Animas River Spill of 2015." Presentation to the Annual Meeting, Southwestern Association of American Geographers. San Antonio, TX, November 5.

McCormick, Peter. 2016. "Watershed Consciousness and the Animas River Spill of 2016." Presentation to the Annual Meeting, American Association of Geographers, San Francisco, CA, April 2.

Nostrand, Richard. 1992. *The Hispano Homeland*. Norman: University of Oklahoma Press.

Olivarius-McAllister, Chase, Mary Shinn, and Shane Benjamin. 2015. "Catastrophe on the Animas—Toxic Water Floods River after EPA Disaster at Gold King Mine in Silverton." *Durango Herald*, August 6. https://durangoherald.com/articles/9779.

Parsons, James. 1985. "On 'Bioregionalism' and 'Watershed Consciousness.'" *Professional Geographer* 37 (1): 1–6.

Potter, James M. 2006. *Animas–La Plata Project*. Phoenix: SWCA Environmental Consultants.

Robertson, David. 2006. *Hard as the Rock Itself*. Boulder: University Press of Colorado.

Rohrabacher, R. Copeland. 1901. "The Great San Juan of New Mexico and Colorado: A Brief History of the Early Days Supplemented by a Review of the Vast Natural Resources of San Juan Country." Durango, CO: *Durango Democrat*.

Schnarch, Samuel. 2016. "The Effect of Superfund on Post-Industrial Communities: An Analysis of Geographic Identity." BA thesis, Program in Environmental Studies, Fort Lewis College, Durango, CO.

Shinn, Mary. 2015. "Utes Bless the Animas River—Southern Ute Sun Dance Chief Hopes to Make It an Annual Tradition." *Durango Herald*, August 16. https://durangoherald.com/articles/1996.

Shinn, Mary. 2018. "Ash Blamed for Fish Kill in Animas River." *Durango Herald*, July 11. htpps://durangoherald.com/articles/231322.

Smith, Duane. 1980. *Rocky Mountain Boomtown*. Boulder: University Press of Colorado.

Thompson, Jonathan. 2018. *River of Lost Souls*. Salt Lake City: Torrey House.

Tuan, Yi Fu. 1974. "Space and Place: Humanistic Perspective." *Progress in Human Geography* 6: 233–246.

Tuan, Yi Fu. 1996. "Space and Place: Humanistic Perspective." In *Human Geography: Essential Anthology*, edited by John Agnew, David Livingstone, and Alisdair Rogers, 444–457. Oxford: Blackwell.

Turner, Jack. 2017. "Durango Re-connect—5th Anniversary a Good Time to Put Aside Differences, Seek Common Ground." *Durango Herald*, September 17. https://durangoherald.com/articles/185026.

US Bureau of Reclamation. n.d. "Animas La Plata Information." http://www.ubr.gov/uc/progract/animas/.

US Department of Energy. n.d. *Durango, Colorado, Processing and Disposal Sites*. Fact Sheet. Washington, DC: US Department of Energy.

6

Tourist Season

LORRAINE L. TAYLOR AND KEITH D. WINCHESTER

The Animas River region felt many impacts from the Gold King Mine spill, and the impact on the region's economy was profound. According to the Leeds School of Business at the University of Colorado Boulder (2018): "The La Plata County economy is highly seasonal and is related to tourism's impact on the local economy and construction. Although significant winter tourism is associated with winter sports, most La Plata County tourism occurs during the summer."

In addition to the magnitude of the environmental and policy concerns, the spill put into perspective the potential for impacts similar events could have on the local economy. Durango typically nears its maximum visitor capacity in early August, so negative impacts on tourism from any event during this time could trigger a ripple effect through a variety of other industries. In addition to tourism spending, the economy also depends on stability in leisure, hospitality, and outdoor recreation positions. Spending in these industries strengthens Durango's economy through the multiplier effect, which sees tourism dollars spent at local businesses. According to the 2018

DOI: 10.5876/9781646421756.c006

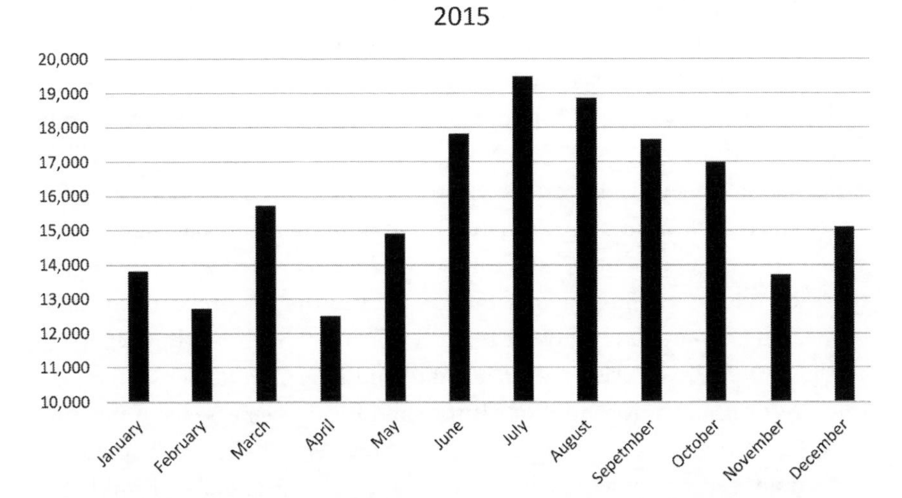

FIGURE 6.1. 2015 enplanements for Durango–La Plata County Airport by month. Revenue Passenger Enplanements 2011–2018. Table provided by Tony Vicari, director of aviation for the Durango–La Plata County Airport on February 7, 2018.

Colorado Business Economic Outlook report, tourism in La Plata County accounts for 20 percent of employment (Leeds School of Business 2018).

TOURISM IN DURANGO

Durango's geographic location makes it the tourism hub of the Four Corners region, and the 18,000 residents welcome approximately 1 million visitors to the area each year (Walsworth 2017). Tourism fluctuates seasonally; the summer season sees visitors coming to visit Mesa Verde National Park (located 35 miles to the west), riding the Durango & Silverton Narrow Gauge Railway, and partaking in a variety of other outdoor activities including biking, hiking, rafting, and fishing. Tourism slows in the spring and fall but spikes in a secondary winter season at the local Purgatory Resort. Monthly enplanement data for the Durango–La Plata County Airport (figure 6.1) provide a metric for analyzing seasonal fluctuations of visitors to the area (Revenue Passenger Enplanements 2011–2018). The data also illustrate the early August date of the Gold King Mine spill occurring during the peak tourism season. The seasonal pattern from 2015 mirrors the few years before and after the spill.

The Durango Area Tourism Office's (DATO) responsibilities as the destination management organization include promoting tourism in and around Durango. DATO purposefully highlights the diverse attractions in its marketing materials; during summer of 2015, its slogan was "A Dozen Vacations in One Destination" (2015). Bob Kunkle, DATO's executive director at the time, logged extensive experience in the tourism industry—specifically in ski resorts—before joining the DATO Board of Directors and eventually stepping into the executive director role in 2014. Kunkle (2018) remembers the wide release of the image of the kayakers as the first indicator of disaster heading toward Durango. He initially felt concern about the ownership of the issue: Whose crisis was it? Kunkle and DATO quickly assumed the role of communication hub for visitors and local businesses and purposefully kept from adopting the problem as their own. The responsibility for crisis management quickly expanded outside of Durango, so Kunkle felt relieved when the ownership of the spill settled above his head. Kunkle and his team at DATO took primary responsibility for communicating with both stakeholders in the tourism industry and current and potential visitors after the Gold King Mine spill. A few of their strategies were particularly successful.

Managing the Tourists

"Durango's Dozen"—the promotion of Durango's multiple tourism possibilities during this time—acted as a blessing during this time, and the abundance of attractions protected the area from an otherwise devastating blow to a summer-dependent tourism economy. Kunkle advised the Durango Welcome Center to answer all calls about the river with suggestions for the many alternative options available for entertainment and activities. Tourists who planned on recreating in the Animas River were pleasantly surprised that the alternatives filled their vacation time, and their Durango experience was otherwise unaffected. Of the perspectives Kunkle monitored at the time, he feels tourists experienced the least emotional connection to the event. Kunkle recalls only one tourist cancellation directly related to closing the Animas. After the river reopened, some anecdotal evidence suggests that tourism may have increased with visitors who came for the novelty or to take pictures.

Managing the Local Residents

Front-line employees played a crucial role in DATO's overall communications strategy. Tourists regularly asked questions of people in positions like hotel front desk agents and restaurant servers. To avoid misinformation in the tourism community, DATO distributed regular updates to tourism businesses, which focused on sharing facts without sensationalizing the event. These messages included tools for front-line employees such as sound bites or quotes gathered from factual sources like the US Forest Service, emergency services, and scientific organizations such as Trout Unlimited. Nearly every guest interaction included questions about the condition of the river, so factual and knowledgeable employee answers became pivotal to DATO's efforts at combating misinformation and negative emotional slants.

Managing the Media

Based on the early distribution and impact of the ubiquitous kayaker photo associated with the spill, Kunkle knew that media coverage posed a significant threat to DATO's image and reputation maintenance efforts. Kunkle partnered with Anne Klein, then employed as a public relations contractor for DATO, to proactively manage the content of the media messages. Kunkle and Klein used a cautious and systematic approach with their public communications language. For example, they proactively limited the use of the name "Durango" in their messaging, anticipating that digital news stories would linger and appear in future tourism-related searches for Durango. They strategically associated the crisis with the mine (the source of the incident) and the river (the vehicle for the mine spill) to protect the Durango brand. Kunkle expressed delight when reporters regularly mispronounced "Animas" in news coverage, creating further discontinuity between the crisis and the Durango area. Their next step involved neutralizing the name of the event itself, and they sent out a series of messages referring to the "Gold King Mine Release" as a purposeful choice to avoid negative language such as "spill," "disaster," or "accident." Kunkle and Klein sought to provide regular and reliable information while simultaneously limiting the potential for lingering brand damage to Durango or its tourism.

Initially, multiple national sources picked up the Associated Press piece about the event (Butler 2015). As with most disaster pieces, national interest in the follow-up was minimal as media sensationalism blew up the disaster but not its recovery. Local and regional media, however, did cover the aftermath, and many of the spill-related news stories and editorials in the *Durango Herald* carried a negative slant, reflecting the local population's ire at the silence and slow responses from the federal level. In addition, the story permeated regional media sources such as the *High Country News* (HCN), which covers stories concerning the American West (Thompson 2015). The author of the HCN piece tempered his story with the awareness of the thousands of other mine-related threats to Western drinking water, reflecting the pique in local curiosity about water sourcing and issues (Thompson 2015). The *Durango Herald* may have published the bulk of articles concerning the spill, and it found aid in distributing them through social media. The *Durango Herald* posted an iconic photograph comparing the pre- and post-spill view of the Animas, which Facebook users have shared thousands of times since August 7, 2015. Social media guaranteed that people connected to Durango found out about the disaster in near-real time, even with huge geographic separation (McBride 2015).

THE TOURISM INDUSTRY

Timing of the Gold King Mine spill was very lucky, in that the summer season was nearly over, and tourism businesses and employees had already reaped the rewards of a solid business. While the river closing crippled certain industries like rafting and fishing, the window was relatively small. Had the spill occurred earlier in the summer, the impacts to tourism could have been much more severe.

Food and Beverage

Dave Woodruff (2017) was the general manager of El Moro Spirits and Tavern Restaurant at the time of the spill. In a town that boasts of having more restaurants per capita than San Francisco (Olivarius-Mcallister 2012), the food and beverage industry is both active and competitive. Despite El Moro's location in the heart of downtown, the majority of its customers are local residents, and their marketing strategy is focused on creating positive word of mouth rather than paid advertising. At the peak of tourism in the summer, Woodruff estimates that tourists make up only 30 percent to 40 percent of El

Moro's business, but he still asserts that "tourism is Durango's largest industry, hands down. It's our lifeblood." Even for a business that doesn't deliberately target tourists, it is aware of tourism's importance: when tourism drops, tourism employees can no longer patronize the business.

This relationship illustrates Woodruff's broader view of tourism in Durango, with each industry supporting numerous others. He also sees food and beverage as the culmination of these relationships, for restaurants provide a middle ground between activities and lodging. As vice president of the local chapter of the Colorado Restaurant Association in 2015, Woodruff witnessed industry heads across the board come together in the incident's wake. Specifically in the food and beverage sector, the initial response involved spur-of-the-moment decisions that inspired Woodruff and other leaders to come together to plan to move forward with a public relations strategy. He credits DATO with providing guidance, keeping communication lines open between key players in town, and sharing the collective voice of the tourism industry during that time.

Woodruff did not recall any cancellations from patrons due to the closing of the Animas River, but he did have a few people approach him who were out of work at rafting companies and were looking to pick up shifts. These were people who had established relationships with many seasonal employees in outdoor recreation and who also had experience in the restaurant industry. There was no decline in sales or staffing for El Moro in the month of August between 2014 and 2015. Woodruff says 2015 was a great fiscal year for most restaurants in town, and the mine spill didn't threaten sales as much as it did morale. Even after the river reopened, it remained a hot topic with customers due to the national media coverage. He was aware of the community relief fund to reimburse lost wages but wasn't aware of any restaurant employees who needed to apply.

Lodging

Kirk Komick (2017) has owned the Rochester Hotel for twenty-five years and is a leader in the local lodging industry. During the summer, the Rochester's visitors are typical of Durango tourists, with a strong interest in outdoor recreation and other active offerings from the Durango Dozen. The Rochester is centrally located in the heart of the historic district and offers complimentary cruiser bikes to its visitors to help them explore amenities in town, like the 7 mile Animas River Trail (2019). However, the hotel's target markets fluctuate throughout the year, and the Rochester is one of the few properties in town that has been successful with the business traveler niche, as well as seeing an increase in international visitors and senior retirees in the fall.

Komick recalls the significance of the iconic photos of the orange river, but his business didn't experience much of a decline and only had calls for two or three cancellations. Komick attributes this to Durango's reputation for having experienced previous pollution impacts. The public perception didn't seem to be affected greatly, as it might have been for a destination known for its pristine environment. His guests at the time seemed compassionate about the impacts of the spill, but it didn't seem to hurt their overall experience or satisfaction with their time in Durango. For him, the timing of the event was lucky, as his summer occupancy historically begins to taper off in August. The Rochester made no adjustments to the room rate due to the mine spill, and business levels for August 2015 set an unprecedented record.

Komick expressed concern that a similar event could happen in the future because of the number of abandoned mines upstream from Durango. Over his twenty-five years in the tourism industry, he has seen demand for river activities increase as offerings cater to more target markets like families instead of attracting only river enthusiasts. The Animas River acts as a significant motivator for his summer visitors. He lamented that the response to the spill by the business community was reactionary, and Komick explained that crisis events are simply "things we can't plan for well." The area had practice during the Missionary Ridge Fire in 2002, a longer-lasting crisis with bigger visible impacts. Again, the media played a significant role in influencing tourists' perceptions and behaviors, especially when the governor of Colorado told reporters that "it looks as if all of Colorado is burning today" (*New York Times* 2002). Komick commended Klein and DATO for their proactive management of the media and efforts to avoid connecting the mine spill with the Durango brand, thereby limiting the negative impacts to his own business and the local economy.

Retail

Antonia Clark's (2017) family has owned the Toh-Atin Gallery since 1983. She has witnessed cycles of economic variations but states that in her experience, "Durango has always been a tourism-driven economy." Over time, retail on Main Avenue has become more tourism-oriented as stores that catered to local residents—like JC Penney, Montgomery Ward, and Woolworths—moved out. Toh-Atin specializes in dealing authentic Native American and Southwest art. Clark estimates that 60 percent of her business comes from tourists, so sales fluctuate with the tourism seasons; a relatively recent surge in online sales also helps balance business throughout the year. In addition to maintaining a profitable business, Clark's mission is to support the weavers and artists as an outlet for their expression and livelihood. According to her, "The most important thing is to keep the weavers weaving."

Clark knows the magnitude of the indirect consequences when an event such as the Gold King Mine spill occurs, with the potential for a decline in sales. The immediate effect on sales was not huge, and Clark recalls August 2015 as "soft." For Toh-Atin, the impact came later: in September, typically the gallery's best month for sales, the numbers were way down from 2014. Clark sees sales tax revenue as both the monitor of a city's health and the key to its survival. It reflects the movement of both tourist and local retail spending, which helps put numbers to the elusive tourism traffic. She identifies recent trends of tourists spending more of their travel budget on food and beverage and less on retail. Local residents appreciate the sales tax brought in by the retail stores on Main Avenue, but since most residents don't shop in those stores, the dependence on tourists' spending is increased.

The lasting impact in the media was a concern for Clark, and she personally called the local newspaper on three occasions to ask them to remove the kayaker photo from the website's homepage. Clark believes the national publicity about the mine spill may have scared away visitors; by November and December, half of Toh-Atin's customers were still asking about it. Clark also harkens back to the Missionary Ridge Fire. In her view, that event differed from the mine spill because of its effects on the quality of life for Durango tourists and residents alike. The smoke obscured visibility, fouled the air quality, and damaged building exteriors. Fire also poses a direct threat to material assets, like real estate and forests. Clark was acutely aware of the precarious state of the multiplier effect during that time because both tourism and local spending plummeted. Those who witnessed the Missionary Ridge Fire in 2002 agreed that it dwarfed the Gold King Mine spill's community impacts.

Rafting

As expected, the businesses that felt the greatest impact were those whose primary functions take place on the river. The executive director of the Durango Chamber of Commerce at the time said, "It was like shutting down Main Street at Christmastime" (Romeo 2016). In Durango during that period of time, there were roughly a half dozen river rafting companies operating guided rafting and kayak trips. Payden Bell (2017) was employed as a guide with Outlaw Tours, which started in 1996, and had achieved trip leader status by that point in the summer. Mid-day on August 5, Bell was guiding a tour when he found out about the mine spill. He knew something serious was taking place when his manager flagged him down from the riverbank and told him to end the trip and disembark his group from the raft before the usual end point. After safely executing this order, Bell started looking up information on his phone and then drove north of town to stand on a bridge and watch the contents of the mine rush into town in a wave of orange sludge.

The rafting companies' immediate strategy involved minimizing negative impacts on their business. The companies strived to maintain their bookings and give their employees work hours. Some, but not all, of the companies offer diversified products, with land-based tours in addition to rafting and kayaking. These businesses worked to accommodate customers with alternative recreational activities such as jeep tours, scenic driving tours, and tours of nearby Mesa Verde National Park to avoid canceling reservations and losing further business.

Historically, rafting companies experience steady business in early August, though by then the water levels have declined significantly from their peak in June. The days of the river closing precisely matched the brief window in the late summer when families are still visiting before their children go back to school. Bell estimated that at that time, he would have been working seven days a week and guiding three to five tours per day. Outlaw offered tours in varying time increments, and customers chose between two-, four-, or six-hour options. Bell made $11 per hour and averaged $20 in tips per trip. By his estimation, his lost wages for the ten days when the river was closed equaled approximately $1,500.

Bell was advised that he had the opportunity to file a claim for lost wages with the US Environmental Protection Agency (EPA). A crisis center was set up at the La Plata County Fairgrounds, and as soon as he was able, he submitted a claim with wage estimates and a recent pay stub. Later that day, he heard about the local recovery fund for wage reimbursements coordinated through Durango's Business Improvement District (BID) organization ("Historic Downtown Durango" 2018). He was advised that he could not apply for the local recovery fund if he had already submitted a claim to the EPA. All of the communication about his options to apply for wage recovery funds was based on word of mouth through his manager and his network of raft-guiding friends. Three months later, he heard from the EPA that his claim had been denied, and he is not aware of any claims that were accepted. Two years after the spill, the EPA contacted him again with an invitation to resubmit his claim. However, he failed to copy his pay stub, which he submitted with the first application. Without a copy of his pay stub as evidence of his lost wages, he was unable to complete the second application.

Bell continued working as a raft guide in the brief period remaining in the 2015 rafting season after the river was reopened. Customers maintained a high degree of awareness of the spill and were curious about the impacts on the river. For these customers, Bell made it a habit to point out rocks on the river that had permanently changed color. Customer questions regarding the spill remained common when he worked as a guide again in 2016, and the level of questioning waned only slightly in 2017. Bell reflected that even international visitors in 2017 were still asking about the spill. He also noted that the talent pool for guides in 2016 was shallow, and his company found it unusually difficult to recruit experienced raft guides in the year following the spill.

As someone raised in La Plata County, Bell feels personally connected to the Animas River. He makes somber predictions about the likelihood that the Animas could experience another mine spill in the future, creating consequences just as dire. While he feels comfortable guiding on the river again without fear about potential health effects, he consciously stays on the surface where the water is moving and no longer goes treasure hunting in the sludge on the bottom.

Fishing

Other riparian-focused businesses include the handful of fishing companies in the community. One of the largest players in this sector is Duranglers, started by Tom Knopick and John Flick in 1983 ("About Duranglers" 2018). By August

2015, the company was employing roughly fifteen guides and a half dozen salespeople who worked in the storefront on Main Avenue selling retail products and booking the tours. Will Hurtgen (2018) was working for Duranglers as a sales associate during the summer of 2015. Hurtgen would answer calls from potential customers, clarify their needs and expectations, and match them with a guide and a location. Customers have a choice of a full day or a half day and of float trips or waded trips. Despite its proximity to the Animas, Duranglers does surprisingly little business on the river and instead books tours on neighboring rivers. Fewer than 10 percent of Duranglers' tours are scheduled on the Animas River. Dependent on snowmelt, the conditions are inconsistent, and only the most experienced customers are taken to the Animas.

The majority of the tours are on the San Juan River, just across the border in New Mexico. The San Juan is considered a Gold Medal River, a title bestowed only on areas that can produce a minimum of twelve quality trout (14+ inches) per acre (Florence 2015). Since it is stocked with trout and has water levels controlled by a dam, the experience for customers booking fly fishing tours on the San Juan is much more consistent than that on the Animas. Duranglers considers the San Juan River to be its bread and butter. Since the river was not affected by the Gold King Mine spill, Duranglers was well positioned to maintain current business levels and continue with planned trips. The concern about river activities in the area, however, was out of the company's control. It soon realized that potential customers misunderstood the services offered, instantly associating the business with the Animas River.

Hurtgen (2018) reflects that Duranglers immediately saw a decline in tour bookings and retail sales because "general demand for river activities went down." Even though the condition of the Animas would have no effect whatsoever on customers' consistently satisfying experience on the San Juan River, he estimated that Duranglers lost 25 percent of its business over the rest of 2015. All of the fifteen guides were impacted in some way, with a reduced number of scheduled tours resulting in lost wages. Retail sales move closely with the tour bookings, creating incentives for satisfied tour customers to make purchases in the shop after returning from a successful trip. Fewer tours resulted in a comparable decrease in retail sales.

Duranglers tried to maintain a media presence as a reassuring voice during the hysteria. The sharp decline in bookings motivated the business to take an educational approach to its external messaging, but it quickly became clear to Hurtgen that the media had little interest in sharing the story, which had been founded on accuracy and reason. Hurtgen perceives that the media blew it out of proportion, which made it impossible for Duranglers to shake its association with the Animas River during and after that time. While redirecting visitors to alternate recreational activities was an opportunity for the rafting companies, fishing companies suffered when customers connected their fear and avoidance of activities on the Animas to all local rivers and river activities.

The timing of the spill was unfortunate for the fly fishing season. While rafting companies had already begun their slow decline into the fall, Hurtgen explained that the fishing companies only begin to ramp up in August. September was historically the busiest and financially strongest month, but with very little momentum coming out of August, Duranglers never climbed to the projected 2015 business levels. Duranglers' long-term concerns focused on local river and fish health, which in turn depends on the resiliency of the insect populations and their recovery from the contamination of their habitat. From its perspective, Duranglers didn't notice any enduring issues with the bug population. Perhaps stimulated by the 2015 visitors who postponed their trips, Duranglers' tour bookings were up year over year in 2016.

ECONOMIC IMPACT

For a city of 18,000, hosting 1 million visitors annually has a considerable economic impact. Residents often express concern for their own quality of life when sharing their town with so many tourists. However, the benefits to the economy are substantial and impact nearly everyone, regardless of whether they are directly or indirectly employed by the tourism industry. The tourism industry is the number-one employer in the city (Clark 2017), and although the service-oriented jobs can be short term with relatively low wages, these positions help sustain other sectors of the economy.

The Gold King Mine event drew attention to the multiplier effect within a tourism-dependent community. Ideally, outsiders bring money into an economy, where it then stays and circulates both within the industry by business owners paying for labor and supplies and outside of the industry by employees buying gas and groceries and similar items. The multiplier effect increases every time the community reinvests the initial amount in itself.

Concern arose immediately over how the spill and the media coverage of the spill would impact the direct spending of tourists at hotels, restaurants, attractions, and souvenir stores. The evidence shows that the potential for an economic crash never came to fruition. In fact, sales tax data indicate that August 2015 was up 2.5 percent from the previous year (Walsworth 2017). Kunkle (2018) explained that there were so many employees of the EPA and the state and federal government that they filled any hotel vacancies. However, visitor spending wasn't the only concern. Many people employed on the river went without hourly wages and gratuities for nearly two weeks (Bell 2017). These people experienced sudden budget reductions, so their discretionary spending on nonessentials was limited. Even though many visitors were spending the money previously allocated for river activities in other areas, the lack of spending from river-dependent employees became a concern in the community.

Rafting companies capable of offering alternative land-based activities did so. This helped maintain income for line-level employees who would otherwise have been raft guides during the ten-day window after the spill. Other companies didn't have the resources to redirect their customers. Rafting and fishing tours were cancelled, and some guides were forced to look for income elsewhere. Many service employees garner experience in various positions in the tourism industry, and newly unemployed raft guides sought to pick up shifts in restaurants or similarly unaffected businesses (Woodruff 2017).

The community also offered financial help to those who could document that they lost wages during the period the river was closed. A relief fund had been created in 2008 after a grease fire at a restaurant closed down an entire block of Main Avenue in the heart of the historic district (Pankratz 2008). At that time, approximately $100,000 was donated to a fund intended to reimburse workers who were unable to earn wages due to the fire (Walsworth 2017). Some money was left in the fund, and additional donations were made after the Gold King Mine spill to reimburse lost wages for employees, mostly raft guides.

Sixty-nine people applied to the Community Emergency Relief Fund and were verified to be employees of rafting companies; the average reimbursement was $500 in the form of gift cards or utility payments (Walsworth 2017). While the financial help from the community was appreciated, full-time raft guides could have been scheduled for multiple trips each day and had the potential to earn $1,500 through hourly wages and gratuities (Bell 2017) during the time the river was closed. These workers, as mentioned, also had a choice of whether to file for reimbursement of their lost wages from the EPA, but all known submissions were denied (Bell 2017).

COMMUNITY LEADERSHIP

The reimbursement fund is just one example of how the business community rallied during the period of the river closure. The La Plata County Economic Development Alliance spearheaded another initiative, which gathered community leaders for sharing information and developing strategies to mitigate the potential economic impact ("Learn about the Alliance" 2018). Leaders from the City of Durango, the Colorado Department of Local Affairs, DATO, the Durango Business Improvement District, the Durango Chamber of Commerce, Four Corners Economic Development, Fort Lewis College, the Region 9 Economic Development District, the Southern Ute Tribe, and others joined the La Plata County Economic Development Alliance to form the Economic and Business Impact Group. Subcommittees were created for regional branding and tourism, as well as economic assistance and recovery. This effort to gather community leaders supported transparent communication, as the information shared in the meetings trickled down to other residents and employees who were impacted.

The regional branding and tourism subcommittee was led by Bob Kunkle of DATO. The subcommittee prepared a report including an analysis of the media coverage, the immediate and long-term business impacts, and recommendations for brand damage recovery (Kunkle 2015). An analysis of the media coverage found that more than 160 articles about the mine spill had been published within twenty days of the event, resulting in nearly 19 million impressions with a cost equivalent to $3.4 million in advertising. Kunkle applied to the EPA for marketing funds to reverse brand damage from the event, but the application was rejected (Pace 2018).

The immediate business impact was designated as "severe" for river-related companies as compared to "mild" for other hospitality-related companies like retail stores, restaurants, and other attractions. Concerns for the long-term business impact were related to lingering media depictions of a contaminated river, negative local chatter by customer-contact employees, and public safety concerns for water and air quality. Intended metrics for the long-term impacts could be a decline in requests for travel planners, advanced reservations, and group bookings. However, DATO's best estimates were that 2016 did not see a decline in demand from visitors as measured by requests for travel planners. Mail requests for hard-copy planners slipped about 3 percent from 37,885 in 2015 to 36,720 in 2016, but downloads of the digital travel planner increased by 75 percent from 2015 to 2016 (Lueck 2018).

The report also made recommendations for proceeding from the brand damage caused by the media coverage and public perception of the spill (Kunkle 2015). Five key areas were identified for recommendations.

The first was to build an arsenal of imagery, intended for accumulating and distributing positive images of the river and other activities in Durango. Tactics to achieve this goal included applying for a $25,000 matching grant from the Colorado Tourism Office to update the photo library and b-roll media inventory, producing several new adaptable promotional videos for social media and tourism websites, using drones for aerial images of the river and large outdoor events, and extending a photo content to the public who could share their high-quality images for use in DATO's promotional material.

The second concern was managing the perception of the Durango brand in search engine results by increasing the quantity and intensity of positive press about the Animas River through releases, photos, packages, and new events. This data dump would dilute the frequency of negative stories appearing at the top of search results about topics such as "Colorado river vacations" and "Durango river rafting."

The third initiative involved coordinating with the city's Parks and Recreation Department in generating media coverage of an event associated with the grand opening of the Animas River Park. In addition to the grand opening event, the fourth recommendation was to increase the impact and media coverage of the annual Animas River Days event, which could result in positive media coverage about the river on a yearly basis.

The fifth and final idea endorsed in the report was working with the Durango Chamber of Commerce in establishing communication channels with front-line tourism employees, providing coaching and tools for talking about this and future incidents with customers at hotels, restaurants, and other attractions.

CONCLUSION

The lessons learned from the Missionary Ridge Fire in 2002 may have positioned Durango to react to and recover from the Gold King Mine spill with relatively minor consequences to the tourism industry and the local economy. The Missionary Ridge Fire consumed nearly 73,000 acres in thirty-nine days and kept Durango citizens on high alert for over a month (Rodebaugh 2012). The detrimental effects included worsened air quality, ash clouds, and property destruction. While fires don't usually fall under federal prevention efforts, this one helped galvanize Durango's emergency protocols. Regardless of state or federal assistance, Durango's local response teams were better prepared for the next disaster.

Immediately following the Gold King Mine spill, the community had more questions than answers. For the most part, locals kept living their lives as before. Granted, some fishing and rafting stopped, but the problems were less tangible than those of a fire. How bad was the damage to the river, to the economy, to Durango's reputation? The accident occurred at the end of rafting season, so displaced raft guides were only a few weeks shy of their natural termination dates. Would they have jobs again next year? These unfulfilled questions hung over the local population's heads and continued to hang as time passed and media coverage decreased. Many business leaders shared the fact that they felt the impacts were much milder than they could have been, but the long-term effects had not yet been measured.

Reflecting now on the Gold King Mine spill, heightened awareness of the dependence of Durango's economy on the tourism industry emerged as a primary outcome. While there was potential for an economic crisis, the economic impact was limited due to a variety of factors that happened to fall in Durango's favor (Elliott 2017). The timing of the spill in early August was less than ideal, but it had a far lower impact on tourists' behavior than a spill earlier in the summer during peak runoff would have had. In a town known for its outdoor recreation, an environmental crisis in June or July could have scared off more tourists who had time to change their summer vacation plans. Initial media coverage surrounding the event felt like an onslaught, but the story faded from headlines relatively quickly, and efforts to dissociate Durango from the crisis successfully minimized brand damage. Community leaders responded quickly and effectively, providing information and financial resources to those in need. Common ties and goals extended across discrete sub-industries, as food and beverage, lodging, retail, rafting, and fishing businesses navigated the crisis with unwavering community support. Kunkle (2018) stated that one good thing that came from the spill was that local community members were reinvigorated to protect their natural resources. Awareness and advocacy raised by this event will help the community's resiliency when it faces similar challenges in the future.

REFERENCES

"About Duranglers." 2018. Duranglers Flies and Supplies. http://duranglers.com/duranglers-fly-shop.

"Animas River Trail." 2019. City of Durango. http://www.durangogov.org/index.aspx?NID=568.

Bell, Payden. 2017. Trip leader and raft guide for Outlaw Tours. Interview, December 20.

Butler, Anne. 2015. "Gold King Mine Spill Captures Attention of World's Media." *Durango Herald*, August 15. https://durangoherald.com/articles/94392.

Clark, Antonia. 2017. Co-owner of Toh-Atin Gallery. Interview, December 21.

"A Dozen Vacations in One Destination." 2015. Press Room for the Durango Area Tourism Office. https://www.durango.org/press-room.

Elliott, Dan. 2017. "Lawyers: Economic Damage from Gold King Mine Spill Less than Thought." *Durango Herald*, April 3. https://durangoherald.com/articles/147990-lawyers-economic-damage-from-gold-king-mine-spill-less-than-thought.

Florence, Jeff. 2015. "Colorado's Gold Medal Waters." Colorado Trout Unlimited. http://coloradotu.org/2015/11/colorados-gold-medal-waters/.

"Historic Downtown Durango." 2018. Durango Business Improvement District. http://www.downtowndurango.org/.

Hurtgen, Will. 2018. Duranglers sales associate. Interview, January 29.

Komick, Kirk. 2017. Co-owner of the Rochester Hotel and the Leland House. Interview, December 21.

Kunkle, Bob. 2015. "Brand Recovery Committee Report." Economic and Business Impact Group. Presented to DATO on September 2 in Durango, CO.

Kunkle, Bob. 2018. Former executive director of the Durango Area Tourism Office. Interview, January 9.

"Learn about the Alliance." 2018. La Plata County Economic Development Alliance. https://yeslpc.com/learn-about-the-alliance/.

Leeds School of Business. 2018. "Colorado Business Economic Outlook." Boulder: Business Research Division of the University of Colorado.

Lueck, Beth. 2018. Digital marketing and website specialist for the Durango Area Tourism Office. Interview, February 1.

McBride, Jerry. 2015. "Before and After." *Durango Herald*. https://durangoherald.com/2015/goldking.

New York Times. 2002. "Fears May Be Outpacing Reality in Colorado Fires." June 16, sec. 1, 12.

Olivarius-Mcallister, Chase. 2012. "Take That, San Francisco." *Durango Herald*, January 17. https://durangoherald.com/articles/33974.

Pace, Jessica. 2015. "How Tourism Officials Calculate Gold King Mine Spill in Ad Dollars." *Durango Herald*, November 26. https://durangoherald.com/articles/98403-how-tourism-officials-calculate-gold-king-mine-spill-in-ad-dollars.

Pankratz, Howard. 2008. "Explosion Injures Durango Firefighters." *Denver Post*, May 7. https://www.denverpost.com/2008/02/22/explosion-injures-durango-firefighters/.

Revenue Passenger Enplanements. 2011–2018. Table provided by Tony Vicari, director of aviation for the Durango–La Plata County Airport, February 7, 2018.

Rodebaugh, Dale. 2012. "39 Days of Destruction." *Durango Herald*, June 7. https://durangoherald.com/articles/39845.

Romeo, Jonathan. 2016. " 'Stigma' from Gold King Mine Spill into the Animas River Could Linger." *Durango Herald*, August 28. https://durangoherald.com/articles/94674.

Thompson, Jonathan. 2015. "Animas River Spill: Only the Latest in 150 Years of Pollution." *High Country News*, August 17. http://www.hcn.org/articles/beleaguered-watershed-animas-spill-epa-durango.

Walsworth, Tim. 2017. Executive director of the Durango Business Improvement District. Interview, December 18.

Woodruff, Dave. 2017. General manager of El Moro Spirits and Tavern Restaurant. Interview, December 21.

7

Contaminated Mines or Minds

The Psychological Reaction to the Animas River Spill

BRIAN L. BURKE, ALANE BROWN, BETTY CARTER DORR, AND MEGAN C. WRONA

When the Animas River was contaminated in August 2015, physical scientists of all types—geoscientists, chemists, and biologists—descended upon the river and the surrounding landscape to collect evidence as to its state of health. As psychologists, however, we take a different view of the situation. Instead of looking for physical clues, we are more interested in people's minds (rather than mines) and their thoughts, feelings, and behaviors in the wake of the environmental disaster.

As such, this chapter frames the psychological reaction to the river spill using the empirical perspectives of ecopsychology, environmental psychology, cognitive psychology, and cross-cultural psychology. Ecopsychology examines people's views of nature and their caring for it, whereas environmental psychology puts forth theories on how people view the physical world, ranging from Goal Frame Theory to Value-Belief-Norm Theory. Cognitive psychology examines heuristics—mental shortcuts—people take to make sense of a highly complex world. The cross-cultural lens adds the important vantage point of the various indigenous groups impacted by the

DOI: 10.5876/9781646421756.c007

spill, such as the Navajo Nation. In putting forward these distinct yet complementary psychological views of the river spill, we hope to shed light on how and why people's reactions to the event differed so dramatically.

ECOPSYCHOLOGY

Ecopsychology addresses the relationship between human beings and their natural environment (Jordan 2009; Milton and Corbett 2011). Theodore Ed Roszak, Mary E. Gomes, and Allen D. Kanner (1995) developed the foundation of ecopsychology by exploring the interconnectedness of the natural environment and the human psyche—damage to one, the natural world, is a direct conduit to damage to the other, human psychological health. Many communities in Colorado, New Mexico, Arizona, and Utah include the natural environment as an important component of their self-identity (Clayton et al. 2013).

On August 5, 2015, various citizens of Durango gathered at the bridge or along the riverbank to watch the plume from the Gold King Mine move down the Animas River and travel through the city. The next morning, television news programs and newspapers showed stark orange/yellow images of the once blue/green river. The images traveled swiftly, and our region was featured on numerous national ("EPA Spill" 2015; Turkewitz 2015) and international news sites ("Millions of Gallons of Waste" 2015). The potentially harmful effects of the 3 million gallons of contaminated mine water flowing through the Animas River that day created an ecopsychological focal point for many impacted communities of the southwestern states (Olivarious-Mcallister et al. 2015).

When our physical environment is damaged, we may internalize that damage (Tidball 2012). In cases of human-made natural disasters such as the river spill, community members may experience a wide range of physical and psychological symptoms (Legerski et al. 2015; Phifer 1990; Phifer and Norris 1989; Quarantelli 2005; Shultz et al. 2015). Because there were no immediate deaths and the damage to the river was likely only temporary, overt symptoms were few in the wake of the Gold King spill. However, the local crisis did undoubtedly harm the communities as a threat to a valued aspect of our inner selves (Doherty and Clayton 2011; Tidball 2012). The process of collective mourning, or *ecogrief*, was articulated in a local newspaper, the

Durango Herald. Numerous photographs of the Animas River following the event were published, and a notable image included a rose and a get-well message on the riverbank that spread throughout Facebook posts and local papers. Coverage of the spill was almost daily, with the *Durango Herald* publishing approximately 275 news reports and updates between August 5, 2015, and August 27, 2016.

Members of the Durango community manifested many of Elisabeth Kübler-Ross's (with Wessler and Aviol 1972) stages of grief in their local press. Following the initial reports in the newspaper, the community responded with *denial* and shock. For instance, many citizens seemed to be surprised that toxic water and heavy metals were predictable components of historical mining communities (Riccard 2015). *Anger* then quickly took shape as various parties and agencies pointed fingers at one another in the media and denied culpability for the event, with irate citizens attacking the US Environmental Protection Agency (EPA) as the chief perpetrator (Marcus et al. 2015; Olivarious-Mcallister 2015; Olivarious-Mcallister et al. 2015; Shinn 2015). Negotiations (*bargaining*) among Silverton, the EPA, Durango, the Navajo Nation, and other impacted agencies and communities then took place.

Sadness (*depression*) permeated the event, with the four most common words repeated among the crowd of river onlookers in the days that followed the spill being "sad," "tragic," "angry," and "toxic" (Thompson 2015). Reopening the river to recreation was a critical step toward healing (Butler and Marcus 2015). Within ten days, the spill of contaminated water from the Gold King Mine was no longer described as a catastrophe (Olivarious-Mcallister et al. 2015)—rather, it was described as a fiasco (Butler and Marcus 2015). It seemed as though the local community had moved toward *acceptance*, the final stage in Kübler-Ross's model. The spill could not be reversed, and it was time to move into healing and recovery.

Ecopsychology also predicts specific patterns of behavior following environmental threats like the river spill (Doherty and Clayton 2011; Tidball 2012). After ecological disasters, environmental action and engagement increase dramatically among locally affected communities (Clayton et al. 2013; Walters et al. 2014). Accordingly, the threat to the river resulted in an increase in conservation efforts and environmental monitoring, particularly among those who self-identified as being harmed (as predicted by Walters et al. 2014). For example, local researchers found a slight increase in the ecological worldview scores of college students after the spill compared to before it, as measured by the New Ecological Paradigm (NEP) scale (Dunlap et al. 2000). The NEP scale is a measure of understanding of ecosystems and their fragility, as well as a commitment to protect endangered environments (Yazzie and Burke 2015). Many citizens internalized the spill as a personal threat and then were motivated to action in efforts of conservation or education (this edited volume is one example).

ENVIRONMENTAL PSYCHOLOGY

To understand our motivation to act, we will examine two prominent theories of environmental psychology that may explain more about people's ecological motivation following the spill. Goal Frame Theory can help us understand how psychological focus, or "goal frame," might have impacted local people's response to an event such as this one (Lindenberg and Steg 2007). According to this perspective, goals fit into three broad categories: hedonic goals, normative goals, and gain goals. The *hedonic* frame—the default for most people—places motivation toward pleasure and enjoyment front and center. The *normative* frame brings attention to our motivation to follow the norms and rules of our social group. The *gain* goal frame focuses a person on their motivation to maintain or improve their own resources.

In the aftermath of the Gold King Mine spill, the normative goal frame—in particular environmental norms—was a major lens through which the people of the Animas River watershed viewed events. Environmental awareness was strong as conversations about the health of the river dominated the media as well as locals' everyday lives. People attended information sessions and followed the unfolding events. Local action groups monitored the river and talked about ways to be involved in the future. A common topic was working to keep this type of accident from occurring again. Thus an environmentalist normative goal frame was firmly in place.

For some people, though, the gain frame was also in play. Tourism is a major source of income in the Animas River watershed, with the river at the heart of this industry. After the spill, city officials "closed" the river, and rafting companies had to suspend operations for two weeks. When the news of the spill hit national and international media, cancellations shot up at hotels. The impact of the drop in tourism rippled out and was felt among many sectors of the local economy. The hedonic frame was operative as well for the many people who used the river or its associated trail system for exercise and recreation.

At any moment, all three of these goal frames—hedonic, normative, and gain—exert some influence. However, if one frame is especially dominant, the others are reduced in relative importance. It has been demonstrated that when hedonic and gain goals are activated, for instance, they reduce the power of the normative frame, which reduces pro-environmental behavior (Lindenberg and Steg 2007). As we watched the joyous river parade after the Animas River reopened, we observed this de-centering of the pro-environmental normative frame. People floated down the river with their dogs balanced on their paddleboards, their beer coolers in their rafts, and their children towing behind on inner tubes. Watchers clapped and river riders gave thumbs-ups and big grins. The hedonic frame was predominant once again. And who could blame us? After weeks of anxiety, here was a triumphant return to pleasure. We could argue, however, based on goal frame theory, that this fun event undermined the pro-environmental normative frame and may have contributed to the continued lessening of environmental concern for the river over the subsequent months and years following the spill.

Another valuable framework utilized by environmental psychologists to understand the presence—or absence—of pro-environmental behavior is the Value-Belief-Norm Theory (VBN) of Environmentalism (Stern 2000). VBN proposes a causal chain leading from values to beliefs to norms and, ultimately, to behavior. Whether the players in the Gold King Mine spill took environmentally protective actions before, during, or after the crisis can be analyzed using the causal sequence of VBN Theory.

The first step in the chain to examine is *values*. A value is a guiding principle in the life of a person or group that spans situations. A variety of taxonomies of values has been put forward by psychological theorists, but VBN focuses on three categories of values that have direct impacts on environmental behavior: egoistic, altruistic, and biospheric values. Egoistic values encompass both the hedonistic goal frame and the gain goal frame discussed above. Sometimes termed egocentric values, they focus on one's own well-being. Examples include someone starting on the path to environmental activism in response to the spill because she is an avid river rafter and wants access to a safe and pristine waterway or someone else participating in a water-monitoring team because he is a fly fishing guide and wants to be able to give firsthand advice to tourists as a way to support his business. If a person is motivated solely by egoistic values, though, environmental values can easily be trumped by alternative choices, such as staying home and watching a movie or improving one's quality of life in ways unrelated (or even antithetical) to protecting the natural environment.

Altruistic values are a similarly unreliable vehicle for environmental action. Looking at the orange water of the plume moving through their hometown, parents may have been motivated to take action to preserve the river for their children. Others may have thought of the ranchers and farmers dependent on the water and felt motivated to do something for the community. Altruistic values are those that uphold fairness, generosity, and kindness toward human beings. However, when the needs of the environment and of people come into conflict, the environment can lose out if altruism is the most central value activated.

According to VBN Theory, then, the most reliable pathway to pro-environmental behavior stems from ascription to biospheric values, which put the natural environment center stage. Also called ecocentric or biocentric values, they value nature for its own sake, as an entity with rights equal to those of human beings. In this view, the Animas River, its surrounding lands, and all the creatures that swim, walk, crawl, and fly in and around it are valued as having a right to exist unmolested. Biocentric values promote an environmentalism that has staying power.

VBN Theory further explains that values provide the framework for *beliefs*. An individual's beliefs normally work together in a consistent network or worldview. One of these that is critical in predicting environmental behavior is the development of an ecological worldview, as measured by the New Ecological Paradigm Scale mentioned above, with local scores increasing after the spill (Yazzie and Burke 2015). An ecological worldview includes an understanding of the concept of intact ecosystems, an understanding of their fragility, the belief that the environment is in peril, and a commitment to protect it. Those who view the world through this lens of belief are more likely to move along the causal chain toward pro-environmental behavior.

The next phase in the sequence of beliefs in the VBN is awareness of consequences. When the orange plume reached Durango, the water looked toxic; then, as the water level dropped, a "bathtub ring" of yellow was left on the banks and on boulders throughout the river. The accuracy of the belief that the mining waste in the river was toxic is addressed elsewhere in this volume as well as further below. Certainly, though, at the time of the crisis, the average person watching the discolored river flow through the community believed that dire consequences were likely.

The next step in the causal sequence of the VBN is the third phase within beliefs: ascription of responsibility. In this phase, we decide whether it is our responsibility to take action. A large literature in social psychology on bystander effects has shown that even if an individual realizes an emergency exists and is alarmed by it, they will not take action unless it seems it is their place to do so (cf. Latané and Darley 1970). Some impediments to passing this stage include the belief that other people's roles or experiences make them more appropriate actors, the presence of many other people, and a desire to avoid the costs of getting involved oneself. As discussed above, various agencies—ranging from the Gold King Mining Company and the EPA to the local and federal government—were part of the puzzle, which could have made many people feel out of their depth as needed contributors to this complex situation.

Nevertheless, some people examined their beliefs and concluded that they did have a responsibility to help the Animas River. At this point, VBN suggests that the *norm* stage occurs. Norms are much more specific structures than values; they govern what is seen to be appropriate or normal behavior. VBN focuses on pro-environmental personal norms, which are the expectations we have of ourselves regarding acceptable behavior in relation to the environment. For instance, many people have internalized personal norms to avoid littering and to recycle whenever possible. In this case, if an individual has the personal norm to get involved when an environmental crisis occurs, they will pass through to the final stage of VBN: behavior.

VBN Theory provides a taxonomy of four categories of pro-environmental *behavior*. The first category is activism, which happened when the local environmental organizations served as watchdogs over the EPA investigation of the spill incident. The second category is non-activist behavior in the public sphere, such as voting for pro-environmental candidates and policies. A third possibility is private-sphere behavior, wherein someone might choose to buy a product made by a manufacturer that donates a portion of the profits to protecting the health of the Animas River. The last is organizational behaviors, such as businesses participating in river cleanup activities and environmental advocacy.

In summary, the VBN Theory of Environmentalism demonstrates how complex the causal sequence is that leads to actual pro-environmental action. A population is more likely to remain environmentally active if it holds biospheric values that inform its view of the Animas watershed, believes that the consequences of the spill are severe, believes it has personal responsibility to do something, and already has norms of action in place. People will then be more likely to hold responsible parties accountable into the future, to personally work toward a healthy river, and to support policies protecting their region from future crises of this kind.

COGNITIVE PSYCHOLOGY

Another perspective on why people responded to the spill the way they did comes to us courtesy of cognitive psychology, the science of how people think and make decisions. From this vantage point, spill reactions can best be understood as stemming from a variety of heuristics or biases—the mental shortcuts humans make in everyday life (heuristics; Simon 1960) and how these mental shortcuts can result in misleading judgments and decisions (biases; Kane and Webster 2013). In this section we will highlight key heuristics that may have contributed to biased public reactions to the river spill, starting with availability and then discussing risk assessment, confirmation and hindsight biases, and the concept of motivated cognition.

Availability is one of the most basic and frequently employed heuristics. This is the process of judging frequency by "the ease with which instances come to mind" (Kahneman 2011, 129). A classic example of availability is that fear of flying is far more common than fear of driving, with lifetime prevalence rates exceeding 6 percent for aviophobia (Depla et al. 2008) compared to just over 1 percent for vehophobia (Becker et al. 2007). Yet driving is clearly more dangerous, with the odds of dying in a motor vehicle accident at 1 in 98 for a lifetime versus 1 in 7,178 for airplanes (Locsin 2008). Other studies have shown that the average number of years you can expect to fly before dying in an air crash is 14,176 (Abbas 2016). Why, then, do people fear more what is in fact less dangerous? Because we overestimate the likelihood of an event where there is greater "availability," or greater "detail," in our memory—and most of us can more readily conjure up detailed mental images of plane rather than car crashes, especially since 9/11 (Abbas 2016).

Similarly, the 2015 Animas River spill produced a highly unusual and therefore durable image in the minds of onlookers—the photo of three kayakers floating in a deep mustard-yellow or orange Animas River near Bakers Bridge that went viral soon after its publication in the *Durango Herald* newspaper, quickly becoming the iconic photo of a mine cleanup gone wrong. It is "a photo that has been seen by countless people and one that defines the enormity of an environmental disaster" (Benjamin 2015). The irony of the impact this and other colored aquatic images had on observers is that what made the river turn yellow/orange—that is, the iron and copper from the mine—is not as toxic as the less visible elements in the contaminated water such as arsenic (William Collins, personal communication, November 2015). In other words, the vividness of the yellow- and orange-colored water made for memorable—but not necessarily accurate—mental images of the dangerousness of the spill, not unlike gory footage of a plane crash.

The *false consensus* effect claims that increased exposure to ideas that confirm our own beliefs will cause us to overestimate the occurrence of similar ideas in others. The ubiquitous coverage of the spill in news articles and online conversations (Butler et al. 2015) likely created an ongoing, reinforcing pattern of ideation that inflated locals' perception of tangible injury from the spill through false consensus. There is no doubt that some harm did occur—particularly financial—but the volume of media coverage and rumination may have exacerbated the perceptions of harm to the typical citizen of the Southwest region (Romeo 2016).

Further muddying the water of our ability to appraise the seriousness of the river spill psychologically is the fact that humans are notoriously weak at *risk assessment* (Eiser 1998). In particular, people tend to overestimate the frequency of rare risks and underestimate the frequency of highly probable ones (Eiser 1998). This is why one Hawaii guidebook advised readers not to worry about shark attacks but to constantly reapply sunscreen—and, of course, one of us went to Maui and got a nasty sunburn but no shark bite. Our minds are therefore already vulnerable to overreact to an event like the river spill and to underplay the much more likely causes of our demise—such as car accidents, heart disease, and cancer.

The *viral effect* is another, more modern heuristic that can lead to biased decision-making; it is the idea that we assume that something is valuable or important only when it gains national or worldwide media airplay. People's appraisal of the dangerousness of the spill was therefore heightened because the spill made national news (e.g., Brumfield 2015). The bias at work here is that people think that if the large news outlets are covering a small-town event, it must be of severe consequence. However, the popularity of a news story does not necessarily result from its importance—see any Kim Kardashian video or tweet for proof. In fact, content that evokes high-arousal positive (awe) or negative (anger or anxiety) emotions is more likely to go viral and be widely disseminated in the media (Berger and Milkman 2012).

So heuristics such as availability and viral media may work in concert with our tendency to misunderstand probability and assess risk inaccurately for rare events like the river spill. But what about people's explanations for the spill itself? One cognitive error many people made in the aftermath of the Animas River spill was hindsight bias (Fischhoff 1975). This is the inclination to see an event as having been easily predictable beforehand, even though it objectively was not (Roese and Vohs 2012). For example, after the conclusion of a sporting event, many people claim they knew who was going to win (Roese and Maniar 1997); it may seem obvious in hindsight that the Denver Broncos' defense would overpower the Carolina Panthers' offense in the 2016 Super Bowl or that the Kansas City Royals would beat the New York Mets in the 2015 World Series. Hindsight bias results from our propensity to "rewrite the story" to make sense of the outcome and focus exclusively on factors we now know are linked to the outcome (Pezzo 2011)—such as Von Miller's strip sacks or Eric Hosmer's superb base running.

Hindsight bias was also evident on August 5, 2015, when the EPA was conducting an investigation of the Gold King Mine near Silverton, Colorado, to treat the water and assess future needs for remediation. While excavating above the old mine tunnel, pressurized water began leaking above the mine tunnel, spilling about 3 million gallons of water stored behind the collapsed material into Cement Creek, a tributary of the Animas River (US Environmental Protection Agency 2016). In the wake of this spill, many people criticized the EPA, claiming it should have known the wall would collapse. Whereas it is obvious in hindsight because the tunnel wall did in fact collapse, it was far from predictable at the time.

Confirmation bias may further help illuminate what happened as criticism of the EPA mounted. Confirmation bias refers to our tendency to seek confirming—rather than disconfirming—evidence for any theory or idea we hold. One consistent example of confirmation bias occurs during political campaigns, when voters tend to look for information to support their preexisting views of the candidates (Westen 2008). Results of brain imaging studies (Westen 2008) in which participants assessed the validity of statements made by political opponents revealed that these unconscious confirmation biases are emotional at their core, likely fueled by fear and threat (Muris et al. 2014): "We did not see any increased activation of the parts of the brain normally engaged during reasoning. What we saw instead was a network of emotion circuits lighting up . . . Essentially, it appears as if partisans twirl the cognitive kaleidoscope until they get the conclusions they want" (Shermer 2006).

After the Animas River spill, people accordingly twirled their cognitive kaleidoscopes looking for evidence to support their own preexisting viewpoint about the government in general and the EPA in particular. If people viewed the government as bumbling, incompetent, or dishonest prior to August 2015, then they were more likely to spin a narrative to explain the spill that involved an overzealous government and the corrupt or inept Environmental Protection Agency (Rebecca Clausen, personal communication, August 2015). Some with even stronger anti-government sentiments viewed the spill as intentional. For instance, the owner of the Gold King Mine claimed he was victimized by the EPA, which had planned the mine spill to usher in Superfund cleanup status (Romeo 2015).

Ironically, these same biases also shed light on why the town of Silverton was against a Superfund designation that may have obviated the need in August 2015 for EPA involvement in the Gold King mining site in the first place. Even after the August 5 spill captured national attention and reinvigorated downstream communities' insistence that the leaky mines be cleaned up, Silverton locals continued to bristle at the suggestion of Superfund (Langlois 2015), although the town did later vote yes in February 2016 (Finley 2016). Two leading reasons for this hesitation were (1) confirmation bias stoking anti-government sentiment and (2) the availability heuristic. Despite the fact that property values actually increase following Superfund cleanup, one longtime Silverton resident put it this way while invoking the availability heuristic: "You hear the word 'Superfund' site and

99 percent think 'danger.' So why would you want to go to a Superfund site" (Langlois 2015).

In this section, we have argued that heuristics and biases explain far more about how and why people reacted to and explained the spill than do logic and reason. In fact, individuals often conform their beliefs about disputed matters of fact—such as why the river spill happened and how dangerous it is—to values that define their cultural identities rather than to statistical or other reasoning per se (Kahan 2013).

CROSS-CULTURAL PSYCHOLOGY

In the case of the Gold King Mine spill, cultural aspects and identities are especially important and impacted people's perception and understanding of the event, as well as the stress-based response (Cwikel et al. 2002). When considering the differential impact of the spill, one particularly relevant concept from the psychological literature is the notion of historical trauma (Brave Heart and DeBruyn 1998). Over time, indigenous individuals in the United States have been subjected to policies that have targeted their rights and ways of living (Dixon et al. 2010). It is not surprising that this invalidation of the cultural experiences of indigenous people has resulted in a strong sense of mistrust between indigenous people and Western culture, including US government entities. This pattern of behavior and treatment has resulted in an ongoing, chronic experience of stress and trauma for an entire population. Similar to an individual who might experience a stress response following exposure to a traumatic event (e.g., increased startle response, avoidance), populations who experience historical trauma may exhibit "symptoms" as well. However, these symptoms often impact the population as a whole through phenomena such as increased suicide rates, addiction, and violence. Elders identify a number of key areas that perpetuate the continued trauma of indigenous people, including the loss of land, loss of language, legacy of boarding schools, ongoing discrimination, and substance use (Grayshield et al. 2015).

Due to this history of trauma, when the Gold King Mine spill dumped toxic material into the Animas and subsequently into the San Juan River, which flows through the Navajo Nation, this brought up memories and thoughts about past traumas and government violations. As has been the case many times in the past, the Navajo Nation now faced a violation of its land and way of life due to the actions of the Western world. For many Navajos, the flow of mine tailings into the San Juan River threatened their relationship with the "water of life" (Romeo 2016). Like other farmers impacted by the spill, Navajo farmers were forced to turn off water sources due to contamination. This took not only an economic toll but also a cultural one, with ceremonial activities adjusted due to the inability to harvest corn pollen, which is used for many traditional rites. In some places, water was not restored for use in farming until one year after the spill, when samples had been tested and deemed safe by a Navajo researcher who could be trusted by the Navajo Nation (Shinn 2016a, 2016b).

The federal government's response following the spill also raised memories of historical invalidation and lack of concern for indigenous people. Whereas representatives from the EPA were quickly present in Silverton and Durango, the EPA was slower to respond to the Navajo Nation—the agency did not initially send a representative to a field hearing specifically related to the impact of the spill on the Navajo Nation and failed to supply immediate disaster relief funding (Romeo 2016). This prompted the Navajo Nation to sue both the federal government and the EPA and perpetuated the legacy of mistrust of government agencies outside of tribes/nations. Although the aftermath of the spill may have revealed that the physical impacts may not have been as devastating as initially thought, the rupture in the tenuous relationship between the Navajo Nation and outside agencies remains, and efforts must be made to rebuild trust that was lost as a result of the spill and the subsequent handling of its aftereffects.

CONCLUSION

In sum, the psychological concepts described above help us understand why we responded to the Gold King Mine spill as we did. Ecopsychology and environmental theories embed our responses in theories of humanity's relationship with nature, whereas cognitive heuristics overlay the flawed functioning of our brains that take shortcuts to making sense of a complicated world. Cross-cultural viewpoints underscore the mounting damage the event and the EPA's slow response might have done to the Navajo Nation's trust of the federal government. In general, understanding the reasons behind our reactions and subsequent actions may enhance our awareness of human response to environmental disasters. The more we can grasp the *why* behind our behavior, the better equipped we are to continue to help our community move toward healing and prevent similar crises in the future.

REFERENCES

Abbas, Ali E. 2016. "Why We Worry about Flying." CNN, May 19. http://www.cnn.com/2016/05/19/opinions/fear-of-flying-misplaced-abbas/.

Becker, Eni S., Mike Rinck, Veneta Türke, Petra Kause, Renee Goodwin, Simon Neumer, and Jürgen Margraf. 2007. "Epidemiology of Specific Phobia Subtypes: Findings from the Dresden Mental Health Study." *European Psychiatry* 22 (2): 69–74.

Benjamin, Shane. 2015 "Durango Kayaker Reacts to Animas River Photo That Went Viral." *Durango Herald*, August 9. http://www.durangoherald.com/article/20150809/NEWS01/150809571/.

Berger, Jonah, and Katherine L. Milkman. 2012. "What Makes Online Content Viral?" *Journal of Marketing Research* 49 (2): 192–205.

Brave Heart, Maria Yellow Horse, and Lemyra M. DeBruyn. 1998. "The American Indian Holocaust: Healing Historical Unresolved Grief." *American Indian and Alaska Native Mental Health Research* 8 (2): 60–82.

Brumfield, Ben. 2015. "By the Numbers: The Massive Toll of the Animas River Spill." CNN, August 13. https://www.cnn.com/2015/08/13/us/animas-river-spill-by-the-numbers/index.html.

Butler, Ann, and Peter Marcus. 2015. "Animas River Reopens after Gold King Mine Fiasco: Health Advisory Issued for Recreational Users." *Durango Herald*, August 15. http://www.durangoherald.com/article/20150814/NEWS01/150819805/.

Butler, Jeffrey V., Paola Giuliano, and Luigi Guiso. 2015. "Trust, Values, and False Consensus." *International Economic Review* 56 (3): 889–915.

Clayton, Susan, Amanda Koehn, and Emily Grover. 2013. "Making Sense of the Senseless: Identity, Justice, and the Framing of Environmental Crises." *Social Justice Research* 26 (3): 301–319.

Cwikel, Julie G., Johan M. Havenaar, and Evelyn J. Bromet. 2002. "Understanding the Psychological and Societal Responses of Individuals, Groups, Authorities, and Media to Toxic Hazards." In *Toxic Turmoil: Psychological and Societal Consequences of Ecological Disasters*, edited by Julie G. Cwikel, Johan M. Havenaar, and Evelyn J. Bromet, 39–65. New York: Kluwer Academic/Plenum.

Depla, Marja F., Margaret L. ten Have, Anton J. van Balkom, and Ron de Graaf. 2008. "Specific Fears and Phobias in the General Population: Results from the Netherlands Mental Health Survey and Incidence Study (NEMESIS)." *Social Psychiatry and Psychiatric Epidemiology* 43: 200–208.

Dixon, Andrea L., Tarrell Awe, and Agahe Portman. 2010. "The Beauty of Being Native: The Nature of Native American and Alaska Native Identity Development." In *Handbook of Multicultural Counseling*, 3rd ed., edited by Joseph G. Ponterotto, J. Manuel Casas, Lisa A. Suzuki, and Charlene M. Alexander, 215–226. Los Angeles: Sage.

Doherty, Thomas J., and Susan Clayton. 2011. "The Psychological Impacts of Global Climate Change." *American Psychologist* 66 (4): 265–276.

Dunlap, Riley E., Kent D. Van Liere, Angela G. Mertig, and Robert Emmet Jones. 2000. "Measuring Endorsement of the New Ecological Paradigm: A Revised NEP Scale." *Journal of Social Issues* 56 (3): 425–442.

Eiser, J. Richard. 1998. "Communication and Interpretation of Risk." *British Medical Bulletin* 54 (4): 779–790.

"EPA Spill Turns Animas River Orange." 2015. CNN, August 8. http://www.cnn.com/2015/08/08/us/gallery/colorado-epa-mine-river-spill-irpt/index.html.

Finley, Bruce. 2016. "Silverton, San Juan County Vote Yes for Superfund Cleanup of Old Mines." *Denver Post*, February 22. http://www.denverpost.com/2016/02/22/silverton-san-juan-county-vote-yes-for-superfund-cleanup-of-old-mines/.

Fischhoff, Baruch. 1975. "Hindsight ≠ Foresight: The Effect of Outcome Knowledge on Judgment under Uncertainty." *Journal of Experimental Psychology: Human Perception and Performance* 104: 288–299.

Grayshield, Lisa, Jeremy J. Rutherford, Sibella B. Salazar, Anita L. Michecoby, and Laura L. Luna. 2015. "Understanding and Healing Historical Trauma: The Perspectives of Native American Elders." *Journal of Mental Health Counseling* 37 (4): 295–307. doi: 10.17744/mehc.37.4.02.

Jordan, Martin. 2009. "Nature and Self—an Ambivalent Attachment?" *Ecopsychology* 1 (1): 26–31.

Kahan, Dan M. 2013. "Ideology, Motivated Reasoning, and Cognitive Reflection." *Judgment and Decision Making* 8: 407–424.

Kahneman, Daniel. 2011. *Thinking, Fast and Slow*. New York: Farrar, Straus, and Giroux.

Kane, Joanne E., and Gregory D. Webster. 2013. "Heuristics and Biases That Help and Hinder Scientists: Toward a Psychology of Scientific Judgment and Decision Making." In *Handbook of the Psychology of Science*, edited by Michael Gorman and Gregory Feist, 437–459. New York: Springer.

Kübler-Ross, Elisabeth, Stanford Wessler, and Louis V. Avioli. 1972. "On Death and Dying." *Journal of the American Medical Association* 221 (2): 174–117.

Langlois, Krista. 2015. "Why Silverton Still Doesn't Want a Superfund Site." *High Country News*, September 3. https://www.hcn.org/articles/why-silverton-still-doesnt-want-a-superfund-site.

Latané, Bibb, and John M. Darley. 1970. *The Unresponsive Bystander: Why Doesn't He Help?* New York: Appleton-Century-Crofts.

Legerski, John-Paul, Andrea Follmer Greenhoot, Eric M. Vernberg, Annette M. La Greca, and Wendy K. Silverman. 2015. "Longitudinal Analysis of Children's Internal States: Language and Posttraumatic Stress Symptoms Following a Natural Disaster." *Applied Cognitive Psychology* 29 (1): 91–103.

Lindenberg, Seigwart, and Linda Steg. 2007. "Normative, Gain, and Hedonic Goal Frames Guiding Environmental Behavior." *Journal of Social Issues* 65 (1): 117–137.

Locsin, Aurelio. 2008. "Is Air Travel Safer than Car Travel?" *USA Today*. http://traveltips.usatoday.com/air-travel-safer-car-travel-1581.html.

Marcus, Peter, Shane Benjamin, and Mary Shinn. 2015. "EPA Takes Blame for Animals River Contamination: Agency Yet to Say How Extensive Damage May Be." *Durango Herald*, August 8. http://www.durangoherald.com/article/20150807/NEWS01/150809708/-1/animasriver/EPA-takes-blame-for-Animas-River-contamination&template=AnimasRiverart.

"Millions of Gallons of Waste Turn River Yellow." 2015. BBC News. http://www.bbc.com/news/world-us-canada-33856444.

Milton, Martin J., and Lisa Corbett. 2011. "Ecopsychology: A Perspective on Trauma." *European Journal of Ecopsychology* 2: 28–47.

Muris, Peter, Suradj Debipersad, and Birgit Mayer. 2014. "Searching for Danger: On the Link between Worry and Threat-Related Confirmation Bias in Children." *Journal of Child and Family Studies* 23 (3): 604–609.

Olivarious-Mcallister, Chase. 2015. "EPA Faces Criticism, Praise at Silverton Meeting." *Durango Herald*, August 11. http://www.durangoherald.com/article/20150810/NEWS01/150819966/.

Olivarious-Mcallister, Chase, Mary Shinn, and Shane Benjamin. 2015. "Catastrophe on the Animas: Toxic Water Floods River after EPA Disaster at Gold King Mine in Silverton." *Durango Herald*, August 8. http://www.durangoherald.com/article/20150806/NEWS01/150809765/-1/AnimasRiver&template=AnimasRiverart.

Pezzo, Mark V. 2011. "Hindsight Bias: A Primer for Motivational Researchers." *Social and Personality Psychology Compass* 5 (9): 665–678.

Phifer, James F. 1990. "Psychological Distress and Somatic Symptoms after Natural Disaster: Differential Vulnerability among Older Adults." *Psychology and Aging* 5 (3): 412–420.

Phifer, James F., and Fran H. Norris. 1989. "Psychological Symptoms in Older Adults Following Natural Disaster: Nature, Timing, Duration, and Course." *Journal of Gerontology* 44 (6): S207–S212.

Quarantelli, Erico L. 2005. "Catastrophes Are Different from Disasters: Some Implications for Crisis Planning and Managing Drawn from Katrina." *Understanding Katrina: Perspectives from the Social Sciences*. http://understandingkatrina.ssrc.org.

Riccard, Nicholas. 2015. "Thousands of Mines with Toxic Waste under the West." *Durango Herald*, August 8. http://www.durangoherald.com/article/20150809/NEWS01/150809596/.

Roese, Neal J., and Sameep D. Maniar. 1997. "Perceptions of Purple: Counterfactual and Hindsight Judgments at Northwestern Wildcats Football Games." *Personality and Social Psychology Bulletin* 23 (12): 1245–1253.

Roese, Neal J., and Kathleen D. Vohs. 2012. "Hindsight Bias." *Perspectives on Psychological Science* 7: 411–426.

Romeo, Jonathan. 2015. "Owner of Gold King Mine Feels Victimized by EPA." *Durango Herald*, November 29. http://www.durangoherald.com/article/20151129/NEWS01/151129571/0/News01/Owner-of-Gold-King-Mine-feels-victimized-by-EPA.

Romeo, Jonathan. 2016. "Navajo Artists Express Experience with Gold Mine Spill." *Durango Herald*, April 11. http://www.durangoherald.com/apps/pbcs.dll/article?aid=/20160410/NEWS01/160419961&tmplate=mobileart&template=mobileart.

Roszak, Theodore Ed, Mary E. Gomes, and Allen D. Kanner. 1995. *Ecopsychology: Restoring the Earth, Healing the Mind*. San Francisco: Sierra Club Books.

Shermer, Michael. 2006. "The Political Brain." *Scientific American*. http://www.scientificamerican.com/article/the-political-brain/.

Shinn, Mary. 2015. "Plans to Plug the Gold King Mine Backfire." *Durango Herald*, August 6. http://www.durangoherald.com/article/20150806/NEWS01/150809720/.

Shinn, Mary. 2016a. "Navajo Farmers Remain Skeptical of River Water." *Durango Herald*, August 6. http://www.durangoherald.com/article/20160805/NEWS01/160809753/0/SEARCH/Navajofarmers-remain-skeptical-of-river-water.

Shinn, Mary. 2016b. "Navajo Researcher Leads Search for Answers about Contamination." *Durango Herald*, August 5. http://www.durangoherald.com/article/20160805/NEWS01/160809754/0/SEARCH/Navajoresearcher-leads-search-for-answers-about-contamination.

Shultz, James M., Lauren Walsh, Dana Rose Garfin, Fiona E. Wilson, and Yuval Neria. 2015. "The 2010 Deepwater Horizon Oil Spill: The Trauma Signature of an Ecological Disaster." *Journal of Behavioral Health Services and Research* 42 (1): 58–76.

Simon, Herbert A. 1960. "Heuristic Problem Solving." In *The New Science of Management Decision*, edited by Herbert A. Simon, 21–34. New York: Harper and Brothers.

Stern, Paul C. 2000. "Toward a Coherent Theory of Environmentally Significant Behavior." *Journal of Social Issues* 56 (3): 407–424.

Thompson, Jonathan. 2015. "When Our River Turned Orange." *High Country News*, August 9. http://www.hcn.org/articles/when-our-river-turned-orange-animas-river-spill.

Tidball, Keith G. 2012. "Urgent Biophilia: Human-Nature Interactions and Biological Attractions in Disaster Resilience." *Ecology and Society* 17 (2): 5.

Turkewitz, Julie. 2015. "Environmental Agency Uncorks Its Own Toxic Water Spill at Colorado Mine." *New York Times*, August 11. http://www.nytimes.com/2015/08/11/us/durango-colorado-mine-spill-environmental-protection-agency.html?_r=0.

US Environmental Protection Agency. 2016. "Frequent Questions Related to Gold King Mine Response." https://www.epa.gov/goldkingmine/frequent-questions-related-gold-king-mine-response.

Walters, A. Brooke, Christopher F. Drescher, Brandy J. Baczwaski, Bethany J. Aiena, Marie C. Darden, Laura R. Johnson, Erin M. Buchanan, and Stefan E. Schulenberg. 2014. "Getting Active in the Gulf: Environmental Attitudes and Action Following Two Mississippi Coastal Disasters." *Social Indicators Research* 118 (2): 919–936.

Westen, Drew. 2008. *The Political Brain: The Role of Emotion in Deciding the Fate of the Nation*. New York: Hachette Book Group.

Yazzie, Jennifer L., and Brian Burke. 2015. "The Effects of Animas River Spill on the NEP Scale." Poster presented at the Undergraduate Research Symposium, Fort Lewis College, Durango, Colorado, fall.

8

Social Impacts of the Gold King Mine Spill on the Animas–San Juan River Watershed Communities

BECKY CLAUSEN, TERESA MONTOYA, KARLETTA CHIEF, STEVEN CHISCHILLY, JANENE YAZZIE, JACK TURNER, LISA MARIE JACOBS, AND ASHLEY MERCHANT

The causes and consequences of environmental disasters reach beyond the biological and physical domain and extend into sociological realms. This chapter begins with a background of the political and economic contexts in which the Gold King Mine (GKM) spill occurred. It then explores how economic priorities mold social relationships to natural resources in the Four Corners region. A literature review follows on how communities respond to environmental disaster; it highlights the divergence between natural disaster and technological disasters. A discussion follows on the role of community-based peer listening models as means to mitigate potential tendencies for technological disasters to disrupt community relationships.

Following the GKM spill, members of the Durango community mobilized to form a loose coalition, called the Animas Community Listening and Empowerment Project (ACLEP). The mission of ACLEP was to provide public spaces for sharing stories of community well-being related to the GKM spill. The ultimate goal of ACLEP is to promote a shared understanding of common values, as well as recognition of differences based on economic,

DOI: 10.5876/9781646421756.c008

social, and cultural histories. ACLEP's work paralleled efforts in neighboring communities along the Animas–San Juan River and on the Navajo Reservation; these included Farmington and Shiprock, New Mexico—the latter of which is located on the reservation. The responses from these public meetings were compiled and are presented here. The strengths and weaknesses of the ACLEP model are then assessed in hopes that it can serve as a guide to inform actions in other communities faced with technological disaster. The chapter concludes by considering the potential creation of a Community Mental Health Emergency Response Plan to address imminent regional environmental disasters in the future.

While other chapters in this volume have explored the geography, history, hydrology, demographics, politics, and economics of the Animas River watershed, less attention has been paid to the ongoing social impacts following the GKM spill. This chapter fills this void by addressing the sociological considerations following the spill, including the role of political economy in the context of environmental harm, community mobilization, and community well-being. The findings presented here result from a process of collaborative research and writing by community members from a variety of occupational and cultural backgrounds. We present an analysis that brings together perspectives from Native American and non-Native researchers, Diné (Navajo) community organizers, social workers, activists, and family/youth advocates.[1] In this way, we hope to continue the development of participatory and decentralized community partnerships that can serve as a foundation for community healing and resilience. Through these partnerships, we hope to create community mental health response plans to better respond to environmental disasters we may collectively face in the future.

SILVERTON, COLORADO: A COMMUNITY CREATED BY A COMMODITY

A common idea among social scientists is that you cannot understand an event without accounting for the social context from which it unfolds. An event like the GKM spill should thus be understood as a consequence of extractive mining in the Rocky Mountain West. After all, acid mine drainage (AMD) spills are not an anomaly but rather a patterned result of a specific social history and ongoing economic relationship with the environment.

The proximate cause of the GKM spill was determined to be the result of a US Environmental Protection Agency (EPA) contractor's miscalculation at the mine's entrance that led to the uncontrolled discharge. However, it can be argued that the ultimate cause of the GKM spill is rooted in the nineteenth century's growth-based industrial economy focused on extracting valuable metals from lands once occupied by indigenous populations. Abundant literature has examined the historical and ongoing linkages between the forces of capitalism, including industrial hardrock mining, and environmental destruction (e.g., Dunbar-Ortiz 2014; Grinde and Johansen 1995; LaDuke 1999; Pasternak 2010; Robbins 2011). This is important to consider in light of the detrimental and ongoing impacts of these relationships in colonial contexts for indigenous populations worldwide (e.g., Carroll 2017; Golub 2014; Kirsch 2014; Wolfe 2006). This historically established structure lays a foundation for the variety of contemporary responses to the GKM spill, each community defined by unique geographic and cultural relationships along the river corridor.

As one enters the headwaters of the Animas River, a sign displayed at the Silverton Visitor Center quickly reveals the social and economic context of this mining region. The welcome sign informs the reader that the town was founded on "silver by the ton." Like many other communities of the late 1800s, Silverton was founded on the basis of a commodity, which propelled westward expansion, colonialism, and industrial capitalism. Economies based on raw commodities (i.e., capital and minerals) create a particular set of social relationships, including the prioritization of market value, the creation of private property, and the externalization of costs. Each is relevant to the ensuing impacts of the GKM spill.

Ecological systems, such as the Animas River watershed, are shaped by social conditions, including norms, traditions, economic rules, and legal arrangements (Dietz et al. 2003). A sociological analysis of the GKM spill emphasizes the growth imperative of capitalism and its historical role of turning nature into commodities (a process referred to as *commodification*). A commodity is a product that is sold on the market and produced for exchange rather than being for personal use. In the production process, human labor interacts with the larger biophysical world either directly or through the use of technology (e.g., machines). While human labor and nature comprise the foundation of all products made for human consumption, the economic system of capitalism prioritizes the market value of commodities (Longo et al.

2015). The social organization of industrial capitalism requires that the commodification process turn aspects of life—both social and ecological—into market-saleable goods and services. However, the commodification of land and labor has particular social and ecological consequences. Political economist Karl Polanyi (2001) called them "fictitious commodities" because the components of life (such as trees, minerals, or animals) were not initially produced to sell on a market. In order to sell them and make a profit, those in power established institutional and legal rules by which nature and the commons could be governed.

The first step in the commodification process was to establish codes whereby communal lands of indigenous populations could be appropriated and transformed into private property of US citizens. This oppressive history has been well documented as the foundation of the United States, and this area of southwest Colorado is no different. The Ute peoples inhabited the large area that extended across present-day Utah, Colorado, and northern New Mexico. Utes migrated to the Four Corners region by 1300 and continued to disperse across Colorado's Rocky Mountains for the next two centuries. In the early nineteenth century, European fur trappers and traders began intruding into Ute territory, and with them came the quest to make profit off of the region's vast resources. Indigenous dispossession of land began with the Treaty of 1849, which forced numerous Western tribes, including the Utes, to submit to the jurisdiction of the United States. This permitted trading posts and military reservations to be established on Native lands. The creation of Colorado Territory in 1861 placed many Utes in distinct jurisdictions, dismantling cultural systems of kinship and migration patterns. Intimidation and violence from the US government led to additional treaties that diminished Ute tribal lands, including the Treaty of 1868, which reduced the land base from 56 million acres to 18 million acres (Potter 2017). Soon after, large mineral deposits were found in the San Juan Mountains, and the Brunot Agreement of 1873 appropriated an additional 3.45 million acres from the Colorado Utes, making way for Anglo settlers to extract wealth through hardrock mining (Decker 2004).

The General Mining Law of 1872 allowed for wholesale mining in the West by giving away federal land to anyone who could show that the land had mining potential. These policies established new social relationships with the landscape, both in the Animas River Basin and throughout the West.

The process to establish mining claims and mining patents was thus available to turn the minerals in the mountains into commodities for capitalist markets. Embedded in this process were the genocide and forced assimilation of indigenous peoples throughout the land.

A mining claim is a parcel of public land defined by four stakes driven into the ground by a mining company or an individual. The federal government has interpreted this right to supersede all other potential uses of public land. A claim does not provide the holder with the right to use the land for any purpose other than mining. A patent, on the other hand, gives the holder outright ownership of mineral-rich land that belongs to the federal government. A patent is a parcel of claimed mineral-rich public land, which the federal government can sell to the claim holder for $2.50 to $5.00 per acre. A mining patent is no longer public land; both the land and the minerals contained in it become private property of the owner(s).[2]

Establishing private property rights allowed the "fictitious commodities" (Polanyi 2001) of the earth's metals to become real, profit-generating market goods. The claim holder was not required to return any money to US taxpayers for the value of minerals extracted. In addition, the environmental "costs" of doing business (pollution, erosion, habitat loss) were externalized to the public, especially those downstream from private mining operations. These basic dynamics of an industrial capitalist economy (commodification, privatization, externalization) shaped the human relationship to the environment, progressively exceeding natural limits and producing ecological, social, and cultural disruptions to processes critical for maintaining ecosystems.

With the many tailings piles created from decades of mining activity, it is not surprising that releases of AMD became a common occurrence. They were so common, in fact, that in 1902, the City of Durango realized that it could not rely on the Animas River for drinking water due to pollution from mining in and around the river's headwaters. Instead, and to this day, Durango obtains its primary drinking water from the Florida River, the next major watershed located to the east of the Animas River watershed. Local Durango historian Duane Smith has stated that Durango residents quickly became aware of the biophysical impacts of polluted river water, including garbage and smelter waste (cited in Romero 2016). However, the social impacts of environmental degradation took much longer to be considered. In fact, it took a toxic spill in Alaska, thousands of miles of away, for sociologists to begin researching and understanding the connection between environmental disasters and community well-being.

SOCIOLOGICAL LESSONS LEARNED FROM THE *EXXON VALDEZ* OIL SPILL (1989)

In the early morning hours of March 24, 1989, the *Exxon Valdez* oil tanker ran hard aground on the Bligh Reef in Prince William Sound, Alaska. The reef's jagged, rocky surface ripped open the hull of *Exxon*'s super-tanker, spilling millions of gallons of crude oil into Prince William Sound, polluting thousands of miles of remote shoreline, and ruining commercial and subsistence fishing grounds. Contingency plans for mitigating an oil disaster were in place but were not taken seriously by the oil industry or effectively enforced by government agencies. Four days would pass as corporations and the US Coast Guard debated who was responsible for cleanup, which delayed critical action. Meanwhile, oil continued to drift down the coast, creating the nation's worst oil disaster at the time.

The commercial fishing village of Cordova, Alaska, with a population of about 2,500 people, was perhaps the community most directly impacted by the *Exxon Valdez* spill. Although the oil didn't touch the shores of the town itself, local fishing grounds were dramatically affected. Fifty-two percent of the town's economy was directly based on commercial fishing, leaving many to wonder how Cordova could survive such a devastating blow. Emotions ran high and viable information ran low.

There was dissension within the community; friends, neighbors, and families were at odds with each other. Disagreements stemmed in part from whether a community member chose to accept paid positions from Exxon to be employed as cleanup workers. There was little faith in government officials, and residents despised industry executives. Incidences of domestic violence spiked, and suicide rates skyrocketed. Families disintegrated. Local businesses shut down. The community failed to band together in the midst of this disaster. In a report documenting the social impacts of the *Exxon Valdez* spill, the Prince William Sound Regional Citizens Advisory Council (2004, viii) concluded:

> Long after the initial response has ended and the local government has returned to routine day-to-day operations, adverse psychological impacts

> associated with disaster continue to erode the social fabric of the community. Results of Exxon Valdez oil spill studies indicate that mental health impacts still persist 10 years post-spill. These impacts include disruption of family structure and unity, family violence, depression, alcoholism, drug abuse and psychological impairment.

The term *corrosive community* was coined by sociologist William Freudenberg (1997), and it captures a set of debilitating social processes first identified in research on how human-caused disasters affect community well-being. As the term implies, the social effects of human-caused disasters damage individuals and communities over a long period of time. These social effects include anxiety, conflict, blame, and stress within community relationships.

Sociologists Duane Gill, Liesel Ritchie, and Steven Picou (2016) followed Cordova's story for over twenty years, conducting the most detailed longitudinal study to date on the impacts of an environmental disaster. A primary finding was the variation in community well-being relative to whether the community experienced environmental harm due to a "natural" or a "technical" disaster. When natural disasters occur, people tend to band together. A flood or tornado may destroy property and infrastructure, but a sense of community typically prevails. Individuals know they are faced with rebuilding and join others in their community to help in cooperative efforts. Natural disasters are defined by a sense of closure, marking an end to the impact. Closure allows the healing process to begin, in what sociologists term a *therapeutic community*. The therapeutic community is explained as exemplifying spontaneous altruism within the group to address the impacts of disaster, in the absence of institutional or organization resources (Tierney 1977). Since natural disasters are often seen as having no one to blame, people feel a sense of connection to help one another and see clear ways to offer assistance.

In contrast, technological disasters (e.g., radiation leaks, chemical spills, food poisoning) offer no predictable pattern that ends with closure. The lack of tangible information around a technological disaster often leaves people uncertain as to what the future holds. People are left wondering, with questions such as "is my water safe to drink" or "will my children's health be affected?" Technological disasters are not predictable and can last for years or even decades without closure. In the absence of closure, mental health issues such as anger, anxiety, and distrust may arise, resulting in the above-described "corrosive community." These socio-emotional impacts can have significant effects on citizens individually, on families, and on the community, creating a secondary trauma of technological disasters (Picou et al. 2004). One of the specific issues the sociologists observed in Cordova was the lack of mental health professionals available to help the community deal with its secondary trauma.

Psychiatrist and medical anthropologist Dr. Spero Manson (Manson et al. 2005) adds critical insight to the concept of corrosive communities, stemming from his extensive research investigating the particular ways mental health issues are manifested in Native American communities. Widely acknowledged as one of the nation's leading authorities on Native health, he demonstrates that the backdrop of generational trauma heightens Native people's susceptibility and potential vulnerability to the consequences of other forms of stressors, such as environmental disaster. Manson and his colleagues (2005, 852) concluded that "American Indians live in pervasively adverse social and physical environments that place them at higher risk than many other Americans of exposure to traumatic experiences." This recognition is crucial, as it demonstrates that variation will exist in community members' responses to environmental disaster based on their social and cultural context.

For example, Lawrence Palinkas and his coauthors (1992) found that six months after the spill, Alaska Natives who lived near the oil spill were characterized by relatively higher levels of depression compared with Natives residing outside the spill region. Ethnographic research on members of the Native Village of Eyak found high levels of subsistence disruption, family disruption, and personal distress in the first years after the *Exxon Valdez* spill (Gill and Picou 1997). This clearly demonstrates that Manson's findings are applicable to the impacts of the *Valdez* spill, causing significant long-term disruptive impacts on the economy, social structure, and subsistence culture of Alaska Natives in Prince William Sound.

The central concept of using peer listening as a community intervention strategy rather than relying on trained professionals emerged through a participatory research model with social scientists, Alaska Natives, and local residents. In June 1995, members of the Native Village of Eyak proposed that a culturally acceptable intervention strategy to address the social impacts of the spill would be to hold a Talking Circle devoted to the *Exxon Valdez* spill. The traditional model of Talking Circles is not unique to Alaska Native culture and is used by Native peoples for a variety of social activities. There are different types of Talking Circles, including Healing Circles, Elder Circles, and Community Circles. Picou (2000, 83) noted:

> The Talking Circle held by the Native Village of Eyak was patterned after the Community Circle. Invitations were posted throughout the Cordova community and sent to the Native Villages of Chenega Bay and Tatitlek. The invitation requested the attendance of "Native people and all those who respect Native culture." The focus of the Talking Circle was the *Exxon Valdez* disaster and the traditional Talking Circle rules of confidentiality and uninterrupted discourse were also noted in the invitation.

The Talking Circle model of the Village of Eyak became a model on which to base community discussion about the oil spill. The two-day Talking Circle event (January 27–28, 1996) allowed community members to share their personal stories and experiences with the impacts of the spill. After the event, the research team learned that the Talking Circle model could help people transform from disaster victims to active participants in their community's survival.

Sociologists studying the *Exxon Valdez* oil spill impacts, in partnership with Alaska Natives, learned that a greater bond of trust could be established with peers than with professionals or strangers. Based on this experience, Picou (2000) extended the model of the Talking Circle to *peer listening circles* as a method for community recovery. Picou recruited peer listening participants who reflected the town's demographic diversity. In a series of workshops, peer listeners were provided with basic crisis response skills, including listen without judgment, validate your peer's experience, and identify as a fellow survivor. Picou continued peer listening training for several years as the town transitioned from a corrosive community into a therapeutic community. People who had spent the previous five years opposing each other were now meeting to determine how they could work together to protect what they valued. The peer listening model proved to be effective in Cordova, and it became a model other communities could use when faced with technological disasters.

Talking Circles, often referred to as Sharing Circles, have emerged from a long cultural history as an innovative indigenous research methodology—

beyond the *Exxon Valdez* case described here. This attempt to decolonize Euro-centric research methodologies has been used successfully in a variety of research settings, including understanding the perspectives of Native American college students (Tachine et al 2016), the meaning of physical activity programs in Native American communities (Lavallée 2009), and how First Nations communities deal with disproportionate driving injuries (Rothe 2009). Rewriting Western research practices is a much needed move in qualitative methodology to provide a culturally appropriate and sensitive method of developing a deep and broad understanding of indigenous participants' verbal descriptions of their feelings, experiences, and modes of reasoning.

Since 1989, technological disasters have increased in frequency and, occasionally, severity. Mining disasters, pipeline spills, and train explosions have become almost commonplace events in our society. While the 2010 British Petroleum/Deepwater Horizon disaster loomed in our collective consciousness, other, simultaneous events like the Enbridge Pipeline spill in Kalamazoo, Michigan, and the Pegasus Pipeline spill in Mayflower, Arkansas, garnered much less attention. The knowledge social scientists have gained from the *Exxon Valdez* case has been applied effectively in other affected communities. This important research on the sociology of environmental disaster, stemming from the inclusion of indigenous practices, proved to be a mobilizing factor in creating community support and analysis after the 2015 Gold King Mine spill in the Animas River.

COMMUNITY MOBILIZATION AFTER THE GOLD KING MINE SPILL

In August 2015, the Gold King Mine (GKM) breached, discharging roughly 3 million gallons of toxic mining waste into the Animas River. A bright orange plume surged downstream—and across social media. The visual impact shocked watershed residents as the sludge drifted through the town of Durango, Colorado, toward Farmington, New Mexico, and then on to the Navajo Nation. In its wake, a sea of emotion and confusion arose. In a coincidental combination of people and events, the prior lessons learned from the Exxon disaster propelled the formation of the Animas Community Listening and Empowerment Project (ACLEP) in the aftermath of the GKM spill.

Immediately after the spill, the EPA held a series of public meetings in Durango to report water quality data and solicit citizen comments. Dr. Becky Clausen, an environmental sociologist at Fort Lewis College, offered public comment about the potential sociological impacts of the spill, referencing the established literature stemming from research related to the *Exxon Valdez* event. Jack D. Turner, a Durango resident with a background in hydrology and social science, was in attendance at the meeting and was also familiar with ACLEP-oriented research. Turner introduced himself to Clausen at the meeting, and they acknowledged that the discussion around the GKM spill had been focused solely on its physical impacts, overlooking the social impacts. Turner became committed to bringing together a loose coalition of community members who shared an interest in understanding the social and emotional impacts of technological disaster.

Using the organizing network of the United Way of Southwest Colorado, Turner disseminated a public announcement for the first meeting, where social impacts of the spill were to be discussed. The meeting was well attended, and ACLEP was born. Lisa Marie Jacobs, a Durango resident, was a key founder of the group. She had lived in Cordova, Alaska, during the Exxon disaster and provided invaluable historical perspective. In addition, Jacobs helped recognize the patterns of a corrosive community beginning to form—anxiety, anger, and blame—in the post-GKM spill context. Other crucial perspectives came from Durango community members including Ashley Merchant (MSW [master of social work]), Scott Smith (from the Mountain Studies Institute), Wendolyne Omaña (family advocate), PTISAQUAH (youth of color advocate), and Anthony Nocella (community facilitator/professor of sociology). The ACLEP project further benefited from collaboration with Diné organizers associated with the community group *Tó Bei Nihi Dziil* (Water Is Our Strength). Members of the grassroots collective included Janene Yazzie, Brandon Benallie, and Teresa Montoya.[3] Dr. Karletta Chief (Department of Soil, Water and Environmental Science, University of Arizona) served as a main collaborator with *Tó Bei Nihi Dziil*. The collective was supported by a social environmental justice grant through the Agnese Nelms Haury Foundation to further involve the Navajo community of Shiprock, New Mexico.[4]

ACLEP made deliberate decisions regarding leadership, decision-making, and funding. Specifically, the group decided to have a horizontal structure, with no hierarchy of leadership; group members also shared facilitation roles. This allowed for engaged participation from a variety of Durango interest groups and created a neutral platform to present to the community (i.e., not representing a specific government entity or nongovernmental organization). Consensus decision-making, rather than a majority vote, was established to consider various perspectives and think through the wide varieties of public opinion that were forming after the spill. ACLEP decided against accepting or applying for any funds or grants, as doing so could potentially influence people's perception of the group's mission or allegiances (INCITE! 2009). A summary of the group's demographics follows:

ACLEP Demographics

Gender: 3 male
- 5 female

Age: 21–35 (4)
- 36–50 (2)
- 50–65 (2)

Ethnicity: 1 Latina
- 1 Native American (Potawatomi, Kickapoo, and Assiniboine)
- 6 white

Occupation: 3 academic
- 5 nonprofit sector

The goal of ACLEP was for river communities to feel empowered by having a means to talk about the GKM spill in a nonjudgmental and safe community setting where participants would feel they were heard. A primary goal of the ACLEP project was to hear diverse perspectives about how community members had been affected and how ACLEP could potentially help facilitate a healing process. It was important to hear from everyday citizens of the various river communities and not solely the EPA, political figures, or media outlets. Based on *narrative therapy* as a foundation, healing starts with first sharing one's story. Narrative therapy involves listening, telling, and retelling how individuals or communities face the challenges or problems they encounter in their lives. This model guided the practices of ACLEP.

Dr. Anthony Nocella, a trained community facilitator and former professor at Fort Lewis College, worked with the group to design a participatory method that could create the space for community listening and empowerment. Out of this process, ACLEP crafted a plan that incorporated the basic tenets of "peer listening" that were suggested by the previous Exxon-related research. In addition, the group worked deliberately to counter the speaker-dominated heritage of Western culture, investing instead in the listener-oriented models of communication that are supported by indigenous cultures (Purdy 1991). The Exxon research showed that peer listening can be a powerful support tool when people do not feel comfortable utilizing the typical mental health networks. The goal is to listen, not to counsel. The focus moves away from remedying the problem to addressing the person.

To this end, ACLEP held multiple public events where community members were asked to provide anonymous written responses to a series of prompts addressing the social and emotional impacts of the GKM spill. The prompts were displayed on large sheets of paper displayed around the room, and participants were given pencils and a Post-It note pad as they entered the facility. The prompts included open-ended questions such as "what does the river mean to you," "how have you been affected by the spill," and "what do you need following the Gold King Mine spill?"

The participants wrote their responses to the prompts on Post-It notes and placed them on the wall under each corresponding prompt. They were given unlimited time, a relaxed environment, and relative anonymity (i.e., no names were attached to responses). This helped to document an authentic range of thoughts and emotions. By publicly posting their responses, participants could also read the voices and stories of their fellow community members. This provided a decentralized and nonjudgmental space to share thoughts and concerns related to community well-being.

In addition to gathering written comments, ACLEP also relied on interviews and the solicitation of artwork to help capture community sentiments on social well-being. ACLEP partnered with local high schools that conducted Story Corps, Inc. interviews in Durango and Silverton.[5] These interviews allowed multiple perspectives to be heard regarding how people were impacted by the spill. The interviews were available for listening by community members at the public event, providing another way to build connections among various perspectives. ACLEP also hosted an art exhibit in Durango to display interpretations by Diné artists, highlighting the various ways Diné communities were affected by the spill. The exhibit, curated by Venaya Yazzie, provided alternative ways of communication between communities to understand how the impacts varied according to cultural, economic, and geographic location.

COMPARING COMMUNITY WELL-BEING ALONG THE RIVER CORRIDOR

A primary goal of ACLEP was to foster better understanding among Animas–San Juan River communities by recognizing the needs and perspectives of people within the watershed as a whole. To this end, a total of six public events were held in three communities along the river corridor, garnering responses from over 125 participants. The first five public events were held in Durango, Colorado, and Farmington, New Mexico. In March 2016, a similar set of prompts was developed by members of *Tó Bei Nihi Dziil* to solicit Diné responses at a community teach-in event at Shiprock, New Mexico, as part of a pilot grant by the Agnese Nelms Haury Foundation, led by the University of Arizona. The residents of Silverton, Colorado, declined to participate in an ACLEP public event. The reason expressed was "meeting fatigue." The small community had been overwhelmed by an unusually high presence of media sources, government officials, and advocacy groups and was averse to further public solicitations.

All written responses from public events were electronically transcribed. Fort Lewis College faculty and student researchers coded the qualitative data that were gathered. This approach fit the analysis well because it gave ACLEP members both a rich source of data and a way to quantify responses from different communities for the sake of comparison.

Common Value, Divergent Impacts

The data gathered from the ACLEP and Diné public events consistently confirmed that the environmental impact of the GKM spill had a social and emotional toll on watershed residents. This in itself is an important finding, as the dominant narrative in media and local government meetings continued to suggest that only biophysical risks and impacts were important to community members. Offering a facilitated time and space to discuss psychological and social concerns opened a new venue for public participation and, we later argue, for empowerment and healing. Two primary themes will be discussed here: (1) the shared perception of intrinsic value among all river community members and (2) the unique ways the social impacts varied among the communities of Durango, Colorado, and Farmington and Shiprock, New Mexico—the last of which is located on the Navajo Nation.

The three communities vary by economic base, state/tribal jurisdiction, geographic features, ethnic diversity, and culture worldviews. One of the only common features among all three entities is that they share the same river corridor. The expectation was that community well-being would be disrupted by the river spill in each location and that economic priorities would be the main cause for concern among the communities' tourism business, farming, and ranching interests. However, ACLEP data painted a different picture. Of the various responses to the question "what does the river mean to you," the most common response in all three communities converged around the theme of the intrinsic value of the river, not economic or recreational values.

Intrinsic value introduces an ethical consideration that the river has value in and of itself, regardless of how it contributes to market and commodity value (i.e., tourism, food production). This is a significant finding because it demonstrates that within the various socioeconomic and demographic differences among the three communities, there is common ground. Common ground is a crucial starting point for community conversations ranging from environmental policy to racial justice and therefore is relevant to future watershed interactions. We want to be clear, however, that although communities may share similar concerns for the intrinsic values of the river, the stakes are quite different due to differences in land ownership and jurisdiction, socioeconomic status, racial demographics, and cultural worldviews.

We found it significant that in each community the word *life* was used most commonly in responses to the questions about the meaning of the river and its intrinsic value. This word needs relevant context to be meaningful; therefore, we explored the data more closely to find out how intrinsic value in each community was most commonly described. We found that in Durango, the intrinsic value of the river was most closely tied to an *individual's* connection to the river. Three representative quotes from Durango residents included:

> Participant 1: "Water is healing to me. Relief from stress. Connection, reflection of ourselves."

Participant 2: "A place that brings me clarity."

Participant 3: "Fresh water and a relaxing place to go."

These quotes represent a common sentiment among Durango responses that the Animas River provides the community with a place for individuals to seek solace, quietness, and contentment. This is in accordance with key research findings that state: "Contact with nature improves psychological health by reducing pre-existing stress levels, enhancing mood, offering both a restorative environment and a protective effect from future stresses" (Gallis 2013, 4). This quote conforms to the Western understanding of "nature," which aligns with the predominant non-Native demographic of Durango—of nature as a concept that has benefits to physical health.

The Farmington community responded with comments of intrinsic value that shared an individual connection to the river similar to that expressed in Durango. However, one variation stood out, in that intrinsic value was also connected to the other parts of the community, including education and the river as a *shared* source:

Participant 4: "It's a precious water source."

Participant 5: "The main source of life, without it, there will be nothing."

Participant 6: "Teaching of Navajo culture. Also the life that brings."

Here we surmise that Farmington's intrinsic values represent more of the collective uses of the river, due to a history of allocating water for multiple uses such as farming, ranching, and mining. In addition, the larger Diné population in the Farmington area introduces the importance of the river as a way to maintain and teach Diné culture.

Finally, the responses from Shiprock community members clearly demonstrated an intrinsic value that stems from a deep spiritual belief system in which the river is embedded. The following quotes from Shiprock residents show intrinsic value in reference to prayer, ceremony, and larger conceptions of space and time:

Participant 7: "Prayer. Ceremony. Holy beings reside here. Life."

Participant 8: "The San Juan represents a male river in Diné culture. To me the river is a life giving force to the Diné people and ecosystems. It is the connector among SW [Southwest] tribes."

Participant 9: "The river means the past of our [Diné] people, the present destruction of our natural world came to us, and the future of our lives here as Navajo people in our own lands we have to protect now or there is no future."

These quotes clearly demonstrate the intrinsic value of the river for Diné members of the Shiprock community, which is different from the individual connection expressed in Durango and the collective recognition of Farmington. The Shiprock responses may show that no clear boundary exists among humans, communities, and rivers.[6]

While the underlying sentiments behind intrinsic value differ for each community, we want to emphasize the important finding that the inherent worth of the river was the prime way participants from Durango, Farmington, and Shiprock expressed their greatest concern. In contrast, more variation was found in the way each community described how it was affected by the spill.

In responding to the prompt "how have you been affected by the spill," all three communities shared a primary theme of emotional impact, which confirmed our expectations regarding the social impacts of environmental disasters. Responses captured sentiments of grief, anger, anxiety, and loss. The interesting variation among Durango, Farmington, and Shiprock was found in each community's second-most-common responses to that prompt. The secondary impacts for Durango corresponded to the theme of recreation; for Farmington, the theme was irrigation/water use, and for Shiprock, it was subsistence use. These impacts correspond to the political economy of each region. This variation must be noted to understand why a "one size fits all" risk assessment or mitigation strategy will not work for the distinct communities. For example, the EPA calculated the threshold limits for concentration of metals in water based on an analysis of a hiker using water for recreational purposes. This is not representative of the Diné residents in Shiprock, who use Animas and San Juan River water for subsistence and spiritual purposes.

The final source of variation in regard to community well-being comes from responses to the prompt "what do you need following the Gold King Mine spill." In Durango, community responses converged around the theme of wanting more information on long-term impacts on the river. There were both explicit and implicit requests for more data in many of the responses to provide information on water quality, sediment contamination, and spring runoff effects. For example, one Durango response that was representative of many stated, "Sediment testing . . . I'm worried about my kids on the beach." Similar requests for more quantitative data were echoed in many of the responses.

In Farmington, responses shared a similar thread of wanting more information about data, as well as ways to work together for prevention and solutions. These responses requested ideas on how to change the policy and legislation to deal with ongoing impacts of AMD. One participant responded that they feel they need "articles about ways to get involved with changing policies"; others commented on wanting to work with "strong-hearted people to protect our only home."

In Shiprock, residents expressed needs that followed a pattern of wanting to discuss prevention and solutions. The comments did not focus on data or policy; rather, they spoke to the role of prayer, the future of Diné people, and empowering sovereign communities. The following response was recorded in Navajo and then translated into English:

> Baa'á'da'hwiilgháago díí doo ánáádoo nííłda'doo. Dooda, ááshoodí! Shidiné'éí bá tsódizin. Tó éí íína áté.
>
> [Let us take care not to let this happen again. Please, do not let this happen again. I pray for my people. Water is life.]

Similarly, another representative response stated the need for "a true water security plan to preserve the future of Diné." These needs expressed by Shiprock community members do not stem solely from the impacts of the GKM spill but rather demonstrate an extension of previous patterns of environmental injustice and historical trauma around resource exploitation on indigenous lands. The next section provides more context for the responses from the Shiprock community.

Environmental Injustice and Historical Trauma

For the several Diné communities who live downstream from the GKM spill, news of the spill blindsided local farmers who soon realized that irrigation water for their crops and animals would be shut off due to the "yellow river."[7] A press release from the time of the spill described how the Diné residents were "scared and felt that they will cease to exist because their main form of sustenance was poisoned" (Turkewitz 2016). The Animas–San Juan River system is a life source for the Diné; it hydrates the land they cultivate and sustains a rich food source for the people. This represents an example of environmental injustice; those who profited from mining and extraction upstream (and, later, mining tourism) place the burden of environmental harm on downstream Diné communities who have been marginalized through colonialism and racial injustices.

The GKM spill halted and disrupted crop production and created financial insecurity for Diné farmers in the Shiprock community. Diné communities depend heavily on the San Juan River for physical, economic, and spiritual activities. Shiprock residents rely on the health of their crops and livestock to provide economic stability as well as yearly income. Chili Yazzie, president of the Shiprock Chapter of the Navajo Nation, estimated that only 20 percent of the farming families replanted in 2016, one year after the spill. He observed that a negative stigma arose after the GKM spill. Other Diné residents living outside Shiprock assumed that all crops grown following the GKM spill would be contaminated. This stigma prevented Shiprock farmers from selling many of the crops that survived the GKM spill (even if they hauled in potable water to keep crops alive during the irrigation shutdown). Yazzie fears this stigma may remain years after they spill because of the continued contamination from prior spill events. He is concerned about this impact on Shiprock farmers and for future generations. This concern is alarming considering the direct correlation with resource loss and psychological distress (Freedy et al. 1994).

Steven Chischilly, a Diné student majoring in environmental studies at Fort Lewis College, conducted a qualitative investigation of the responses from Shiprock residents in the context of historical trauma and environmental injustice.[8] The ongoing ecological degradation of indigenous lands, such as radioactive contamination caused by extensive uranium mining on the Navajo Nation (Voyles 2015), continues to reproduce experiences of historical trauma in several Diné communities. Based on this frequent problem, Chischilly became interested to see if similar patterns were evident after the GKM spill.

From his analysis of Shiprock community responses, Chischilly found three emerging themes: (1) an expressed loss of spiritual connection to the river, (2) concerns of abandonment, and (3) the perpetual experience and infliction of intergenerational trauma as a result of the GKM spill. For example, many expressed concerns about health threats to their water supply and the possible effects the spill could have on their ability to practice particular ceremonies. Chischilly (2017, 11) stated: "The spill not only threatened their water supply, but it threatened the entirety of their being. The river is a direct connection with the divine beings of their culture. In many instances, this act could be classified in part as deicide; the killing of a deity."

Likewise, other responses shared during the Shiprock Teach-In indicated feelings of trauma in the form of neglect or oversight. This contributed to a sense of "expendability" among some residents. Moreover, other responders expressed that they felt abandoned not only by the federal government, reflecting a patterned structure of "broken promises," but by society as a whole. This erosion of trust is also evident in the fact that the Navajo Nation is a plaintiff in a lawsuit against the EPA.

In some instances, Native American participants in the Durango community commented that they do not accept the legitimacy of water quality data presented by government and state officials. This raises the question of the ability of government officials to effectively communicate results to Diné and other regional tribal communities. Moreover, this reflects a larger federal regulatory challenge to adequately consider risks for Diné "river users," in light of their unique interactions with the San Juan River for spiritual, cultural, farming, and ranching activities. Diné responses from the Durango community spoke of needing "indigenous perspectives, not just the scientific EPA info" and "awareness and support of Native American opinion." For over five centuries, the governing settler cultures (i.e., Anglo-American) have colonized land and culture; often, the intentional and deliberate destruction of the Native American nations has been the result. It is not surprising that the reaction to an environmental disaster created feelings of distrust and anger from people indigenous to the San Juan River corridor (i.e., Diné, Ute, Jicarilla Apache).

Despite these feelings of mistrust, Chischilly's analysis also documented responses from the Shiprock community that demonstrated instances of perseverance and resilience—or the dynamic ability to recover from severe disruption. The comments focused on mitigation and future community education. Chischilly (2017) suggests that the optimistic responses show that in spite of the many problems associated with living in a community affected by water contamination, some members of the Shiprock community continue to believe they will see a new, abundant resource one day.

EVALUATING THE ACLEP MODEL

The last section of this chapter evaluates the ACLEP model of community listening by addressing strengths, weaknesses, and plans for future action. The existing literature documents the benefits of community listening, and the ACLEP model extends these in two important ways by including partnering with youth and tribal entities. This listening model shares attributes with the existing Diné model of listening that is exercised at the government level (e.g., chapter house meetings), spiritual realms (e.g., ceremonies), family life (e.g., home—elders, adults, and children interacting and teaching), and healing/health (e.g., Talking Circles). When state and federal government agencies came to address the GKM spill, they did not understand this culture of listening. *Tó Bei Nihi Dziil* (the Diné community organization—TBND) has done an excellent job of facilitating listening sessions along with discussion and action. Much has been learned from TBND's leadership and community organizing.

Partnering with schools in the Animas–San Juan basin was critical and important to the success of ACLEP. We initiated contact and follow-up activities with Animas High School, Piedra Vista High School, Navajo Preparatory Academy, Fort Lewis College, and Southwest Colorado Community College. Schools provide a natural forum for community dialogue; with learning the institutional norm, the GKM spill provided a platform for learning. While local governments, health organizations, and nonprofits focus on the physical science of the spill, schools have a breadth of interests and creativeness with which to approach a topic—as demonstrated in this current work. Clearly, schools approached the topic from the perspective of the physical sciences but also through the expression of art and music. At a public meeting in Farmington, New Mexico, ACLEP partnered with two high schools, and the event included a collaborative art project (batik-print banners) between two schools and a spoken work poetry performance. These youth-initiated projects provided additional dimensions that speak to community well-being.

At another public listening event, ACLEP partnered with a sociology class at Animas High School in Durango. The sociology class had begun its study of the GKM spill by interviewing citizens throughout the watershed using the Story Corps, Inc. platform. This provided an opportunity for others both in and outside the community to hear what fellow and neighboring citizens were saying about how they were impacted by the spill. These interviews were exhibited during an ACLEP public event, which brought in parents to participate by being interviewed, informed, and heard.

The ACLEP model was also unique in its opportunity to partner with neighboring tribal entities that were embarking on their own community research projects. TBND organized and sponsored the Shiprock Teach-In; the group rose up in response to environmental injustice. TBND, led in part by Janene Yazzie, sponsored teach-ins about water sovereignty and the GKM spill.

In a parallel effort, Diné researchers at the University of Arizona (UA), led by Dr. Karletta Chief, were working to address the concerns communities were raising about the GKM spill that were not being answered by government entities. The UA and TBND partnered to apply for funding to continue their research collaboration. Through the aforementioned grant awarded by the Agnese Nelms Haury Foundation, the team was able to bring together Diné scientists and Diné community leaders to discuss work related to the GKM spill in a forum so community members could learn about ongoing efforts to understand environmental, public health, and social impacts of the GKM spill. ACLEP was invited to the teach-in to report its findings from Durango and Farmington, with the intent of sharing perspectives from along the river corridor. The ACLEP model was not new to the Shiprock Chapter community. Various forms of listening sessions are often employed at Shiprock Chapter meetings; they correspond to traditional cultural practices. The ACLEP model added an additional component of asking people to write their thoughts on paper, possibly leading to more interaction and the sense that their voice was heard. A long-term goal of listening sessions is to engage with people in the direct process of making a change.

The ACLEP endeavor also served as a mechanism to bring together researchers from regional institutions of higher learning. Through her community involvement, Dr. Becky Clausen from Fort Lewis College was introduced to UA's Dr. Karletta Chief, who was initiating an in-depth research project on the environmental and social impacts of the GKM spill on the Navajo Nation. The potential for collaborative research demonstrates how community listening efforts can translate into lasting social relationships.

Weaknesses

The ACLEP model was informative and inclusive, and it had significant findings. It must be clear, however, that these are preliminary findings based on informal public gatherings. The ACLEP public forums were not representative of entire communities and were not organized according to specific sampling techniques or rigorous study design. Rather, they were publicly advertised and relied on whoever was able and willing to participate. Future work relating to the listening and empowerment project could be performed on a larger scale, over longer time periods. ACLEP would also benefit from more structure to report results to each community. ACLEP's initial plan to have an interactive website to share results was not implemented due to funding and staffing limitations.

Community Mental Health Emergency Response Plan

A long-term goal of ACLEP will be advocating for a Community Mental Health Emergency Response Plan for the Animas–San Juan River basin. Environmental disasters have dire consequences for local communities when secondary trauma is ignored, as described above. According to Ronald Kessler and his colleagues (2012, 36), "Research suggests that natural disasters can lead to increased population prevalence of mental illness in the range from 5% to 40%." O. Lee McGabe and his coauthors (2014) found that the urgency of being prepared for secondary trauma after a major disaster has been regarded as an important step in addressing "public health emergency preparedness," involving the gathering of numerous varied people in the private and public sectors to be part of the process of healing.

While La Plata and surrounding counties have both the nonprofit, tax-exempt Axis Health System and private practices to help with post-disaster mental health problems, the surrounding communities would be hard-pressed to effectively handle the array of mental- and health-related issues following a large-scale natural or technical disaster. McGabe and coauthors (2014, 511) indicate that "psychological casualties can outnumber physical injuries by substantial ratios and often overwhelm existing clinical resources, particularly in low-resource settings." This shows that there is evidence that people's mental health should be a consideration for community planning and program implementation to address if a disaster ever happens again.

Going forward, ACLEP's work will focus on enabling communities to have a well-run mental health response plan along with their traditional emergency response plans to take immediate action to confront the emotional and mental effects of a disaster. For example, after the *Exxon Valdez* oil spill, the Valdez, Alaska, community came together to create *Coping with Technical Disasters: A User Friendly Guidebook* (Prince William Sound Regional Citizens Advisory Council 2004). Communities in the Four Corners region should learn from such available templates and tailor a plan that fits the needs of this area's bioregion and demographics. Certainly, there is no one-size-fits-all approach, especially given the unique social, economic, historical, and cultural contexts of the Animas–San Juan River basin Yet local communities should proceed by reviewing the sets of guidelines that have been drafted previously and begin to alter them based on the unique needs of the people and places across our region. We admire the dedicated effort Diné researchers from Northern Arizona University (NAU) and UA have put toward gaining funding for a Community Response Plan for environmental disasters on the Navajo Nation, and ACLEP offers to support their work in any way possible.

CONCLUSION

There is an old chemist's adage that the ultimate solution to pollution is dilution; add more water to the pollution and it will dilute it enough to make the downstream water less polluted, thereby lessening the impacts. The public responses from the ACLEP listening sessions challenged the linearity of the physical sciences. As the polluted water passed through communities and the water itself returned to "normal" water quality conditions upstream, the problems became progressively worse for the communities farthest downstream. ACLEP helped reveal that the chemist's adage does not hold true for coupled human-natural systems. Community impact will vary greatly, in ways not directly related to geography and linearity. Rather, the community impact will depend on the social context in which the environmental disaster unfolds, including the economic, cultural, and historical conditions.

Paying attention to the well-being of the community is as important as studying the physical data of an environmental disaster. Conditions do not return to "normal" when normal no longer exists, whether due to the impacts of colonization, economic exploitation, or continued environmental degradation. Therein lies our new understanding of community health, one in which economic, environmental, sociological, psychological, and spiritual conditions are equally considered alongside the physical data and their interpretation.

NOTES

1. The name *Navajo* has its origins in the period of Spanish colonization in what is presently called New Mexico. In this chapter, the preferred term of self-identification, *Diné*, will be used to describe individuals accordingly. The Navajo Nation is the official name of the sovereign entity of which Diné are members. Therefore, the name *Navajo Nation* will be used. For a history of the "Navajo" name, see Schaafsma (2002). For a discussion of broader Diné historiography, see Denetdale (2007).

2. Since October 1, 1994, the Bureau of Land Management (BLM) has been prohibited by acts of the US Congress from accepting any new mineral patent applications. The moratorium has been renewed annually through the various Interior Appropriations Acts. It is unknown how long this moratorium will continue.

3. Teresa Montoya is a PhD candidate in the Department of Anthropology at New York University. Her dissertation research engages issues of jurisdiction, environmental contamination, and political action for Diné communities in and around the Navajo Nation.

4. http://www.haury.arizona.edu.

5. https://storycorps.org/about/.

6. The interpretations of the Shiprock responses are initial observations that are lacking the depth of a detailed analysis, with Diné cultural expertise. We make no attempt to claim understanding of the complex belief systems of Diné culture. Given the exploitative history of social scientists making broad claims about indigenous culture, the coauthors want to acknowledge the fact that we had limited access to the community and limited understanding from being positioned as non-Native researchers working with Native community organizers.

7. This section of the chapter was written by Steven Chischilly, a Diné student at Fort Lewis College, as part of his Environmental Studies Senior Research Project.

8. Historical trauma is defined as the cumulative emotional and psychological wounds that are carried across generations from traumatic experiences such as forced relocation, genocide, and forced assimilation through boarding schools (e.g., Brave Heart 2000; Brave Heart and DeBruyn 1998).

REFERENCES

Brave Heart, Maria Yellow Horse. 2000. "Wakiksuyapi: Carrying the Historical Trauma of the Lakota." *Tulane Studies in Social Welfare* 21: 245–266.

Brave Heart, Maria Yellow Horse, and Lemyra DeBruyn. 1998. "The American Indian Holocaust: Healing Historical Unresolved Grief." *American Indian and Alaska Native Mental Health Research* 8: 56–78.

Carroll, Clint. 2017. *Commentary: The Environmental Anthropology of Settler Colonialism, Part II*. Anthropology and Environment Society blog, May 16. https://aesengagement.wordpress.com//05/16/commentary-the-environmental-anthropology-of-settler-colonialism-part-ii/.

Chischilly, Steven. 2017. "The Socio-Emotional Impacts of the Gold King Mine Spill on the Navajo People of Shiprock, NM, Navajo Nation." Environmental Studies Senior Seminar Final Research Project, Fort Lewis College, Durango, CO.

Decker, Peter. 2004. *The Utes Must Go! American Expansion and the Removal of a People*. Golden, CO: Fulcrum.

Denetdale, Jennifer. 2007. *Reclaiming Diné History: The Legacies of Navajo Chief Manuelito and Juanita*. Tucson: University of Arizona Press.

Dietz, Thomas, Elinor Ostrom, and Paul Stern. 2003. "The Struggle to Govern the Commons." *Science* 302: 1907–1912.

Dunbar-Ortiz, Roxanne. 2014. *An Indigenous Peoples' History of the United States*. Boston: Beacon.

Freedy, John, Michael Saladin, Dean Kilpatrick, Heidi Resnick, and Benjamin Saunders. 1994. "Understanding Acute Psychological Distress Following Natural Disaster." *Journal of Traumatic Stress* 7: 257–273.

Freudenberg, William. 1997. "Contamination, Corrosion, and the Social Order." *Current Sociology* 45: 19–39.

Gallis, Christos. 2013. *Green Care: For Human Therapy, Social Innovation, Rural Economy, and Education*. New York: Nova Science.

Gill, Duane A., and J. Steven Picou. 1997. "The Day the Water Died: Cultural Impacts of the *Exxon Valdez* Oil Spill." In *The* Exxon Valdez *Disaster: Readings on a Modern Social Problem*, edited by J. Steven Picou, Duane A. Gill. and Maurie J. Cohen, 167–192. Dubuque, IA: Kendall-Hunt.

Gill, Duane A., Liesel Ritchie, and Steven Picou. 2016. "Sociocultural and Psychosocial Impacts of the Exxon Valdez Oil Spill: Twenty-four Years of Research in Cordova, AK." *Extractive Industries and Society* 3: 1105–1116.

Golub, Alex. 2014. *Leviathans at the Gold Mine: Creating Indigenous and Corporate Actors in Papua New Guinea*. Durham, NC: Duke University Press.

Grinde, Donald, and Bruce Johansen. 1995. *Ecocide of Native America: Destruction of Indian Lands and Peoples*. Santa Fe, NM: Clear Light.

INCITE! Women of Color against Violence. 2009. *The Revolution Will Not Be Funded: Beyond the Non-Profit Industrial Complex*. Durham, NC: Duke University Press.

Kessler, Ronald, Katie Mclaughlin, Karestan Koenen, Maria Petukhova, and Edward Hill. 2012. "The Importance of Secondary Trauma Exposure for Post-Disaster Mental Disorder." *Epidemiology and Psychiatric Sciences* 21: 35–45.

Kirsch, Stuart. 2014. *Mining Capitalism: The Relationship between Corporations and Their Critics*. Berkeley: University of California Press.

LaDuke, Winona. 1999. *All Our Relations: Native Struggles for Land and Life*. Cambridge, MA: South End.

Lavallée, Lynn. 2009. "Practical Application of an Indigenous Research Framework and Two Qualitative Indigenous Research Methods: Sharing Circles and Anishinaabe Symbol-Based Reflection." *International Journal of Qualitative Methods* 8: 21–40.

Longo, Stefano, Rebecca Clausen, and Brett Clark. 2015. *The Tragedy of the Commodity: Oceans, Fisheries, and Aquaculture*. New Brunswick, NJ: Rutgers University Press.

Manson, Spero, Janette Beals, Suzell Klein, and Calvin Croy. 2005. "Social Epidemiology of Trauma among Two American Indian Reservation Populations." *American Journal of Public Health* 95: 851–859.

McCabe, O. Lee, Natalie Semon, Carol Thompson, Jeffrey Lating, George Everly, Charlene Perry, Suzanne Moore, Adrian Mosley, and Jonathan Links. 2014. "Building a National Model of Public Mental Health Preparedness and Community Resilience: Validation of a Dual-Intervention, Systems-Based Approach." *Disaster Medicine and Public Health Preparedness* 8: 511–526.

Palinkas, Lawrence, John Russell, Michael Downs, and John Peterson. 1992. "Ethnic Differences in Stress, Coping, and Depressive Symptoms after the Exxon Valdez Oil Spill." *Journal of Nervous and Mental Disease* 180: 287–295.

Pasternak, Judy. 2010. *Yellow Dirt: An American Story of a Poisoned Land and a People Betrayed*. New York: Free Press.

Picou, J. Steven. 2000. "The 'Talking Circle' as Sociological Practice: Cultural Transformation of Chronic Disaster Impacts." *Sociological Practice* 2: 77–97.

Picou, J. Steven, Brent Marshall, and Duane Gill. 2004. "Disaster, Mitigation, and the Corrosive Community." *Social Forces* 82: 1493–1522.

Polyani, Karl. 2001. *The Great Transformation: The Political and Economic Origins of Our Time*. Boston: Beacon.

Potter, James. 2017. "Ute History and the Ute Mountain Ute Tribe." *Colorado Encyclopedia*. http://coloradoencyclopedia.org/article/ute-history-and-ute-mountain-ute-tribe.

Prince William Sound Regional Citizens Advisory Council. 2004. *Coping with Technological Disaster: A User Friendly Guide Book*. Anchorage, AK: Prince William Sound Regional Citizens Advisory Council.

Purdy, Michael. 1991. "Listening and Community: The Role of Listening in Community Formation." *Journal of the International Listening Association* 5: 51–67.

Robbins, Paul. 2011. *Political Ecology: A Critical Introduction*, 2nd ed. Hoboken, NJ: Wiley Blackwell.

Romero, Jonathan. 2016. "Durango Will Get Its Water This Winter Solely from the Animas River." *Durango Herald*, November 17.

Rothe, John. 2009. "Innovation in Qualitative Interviews: 'Sharing Circles' in a First Nations Community." *Injury Prevention* 15: 334–341.

Schaafsma, Curtis F. 2002. *Apaches De Navajo: Seventeenth-Century Navajos in the Chama Valley of New Mexico*. Salt Lake City: University of Utah Press.

Tachine, Amanda, Eliza Yellow Bird, and Nolan Cabrera. 2016. "Sharing Circles: An Indigenous Methodological Approach for Researching with Groups of Indigenous Peoples." *International Review of Qualitative Research* 9: 277–295.

Turkewitz, Julie. 2016. "Navajo Nation Sues E.P.A in Poisoning of a Colorado River." *New York Times*, August 16.

Voyles, Traci. 2015. *Wastelanding: Legacies of Uranium Mining in Navajo Country*. Minneapolis: University of Minnesota Press.

Wolfe, Patrick. 2006. "Settler Colonialism and the Elimination of the Native." *Journal of Genocide Research* 8: 387–409.

9

The Problems of Litigating Hardrock Mining

MICHAEL A. DICHIO

On August 5, 2015, the US Environmental Protection Agency (EPA)—attempting to clean the Gold King Mine in Silverton, Colorado—inadvertently triggered the release of over 3 million gallons of contaminated water into Cement Creek, which feeds into the 162-mile Animas River. As the plume of heavy metal–laden, acidic orange water moved downstream into the San Juan River, it brought national attention to Durango and Silverton and the environmental issues surrounding the number of abandoned or inactive mines in Colorado. Ultimately, the disaster affected waterways in Colorado, Utah, New Mexico, and the Navajo Nation.

The disaster that communities along the Animas River witnessed will almost certainly happen again in various places in the West. Abandoned mines are a serious issue that has gone largely overlooked nationally. The US Bureau of Land Management estimates that there are nearly 500,000 abandoned mines across the United States. In the West, there are roughly 161,000 abandoned hardrock mines across twelve Western states and Alaska; 33,000 of them reportedly leak contaminants into nearby lands and waterways

DOI: 10.5876/9781646421756.c009

("Gold King Mine Spill" 2016). More than that, the EPA estimates that the leakage from these mines impacts 40 percent of Western waterways. In Colorado alone, state health officials report that 230 identified abandoned mines have contaminated about 1,645 miles of state waterways ("Gold King Mine Spill" 2016).

It is difficult for government or government agencies to force cleanup of abandoned un-owned mines. Citizen suit provisions, contained in the Clean Water Act (1972) and the Comprehensive Environmental Response, Compensation, and Liability Act (CERCLA; better known as the "Superfund" law, 1980), have become a very common way to deal with issues of polluted waterways. Essentially, these suits allow ordinary citizens to sue for enforcement of the laws without having to demonstrate personal injury in the same manner as normally required in non-environmental tort suits (Tuhus-Dubrow 20015, 153).

While these citizen suits offer a partial solution to enforce EPA regulations and federal law, they are by no means the best approach to resolving an issue this widespread. Instead, citizen-initiated litigation—often termed private enforcement litigation (Farhang 2010, 3–4)—individualizes conflict rather than considering broader, national solutions to an issue (Melnick 1983). The judiciary and litigation are useful only when the executive and legislative branches cannot resolve these abandoned mine environmental issues.

Given the widespread existence of abandoned mines, other visible disasters like the Gold King Mine spill will continue to occur. Consequently, the purpose of this chapter—and its place in this edited volume—is to use the Animas River spill as a starting point for a broader discussion about the litigation engendered by the Clean Water Act and how this tactic, because of its individualizing effect, is not the best way to prevent these disasters from happening in the future. More widespread reform, especially changes concerning legal liability and creating funds for cleanup, would prove more effective. The Good Samaritan Cleanup of Orphan Mines Act[1] and the Hardrock Mining and Reclamation Act[2] are some of the better solutions put forth thus far, recognizing the issues over legal liability for cleanup and the lack of funding. In particular, the requirement that mining companies or private environmental groups take on responsibility for others' previous damages is hindering cleanup, not helping it. The Superfund and the Clean Water Acts are keeping both mining companies and state governments from active reclamation of abandoned sites; thus the threat of another Gold King Mine spill will continue to loom over the West. Significant policy changes are critical to decrease this likelihood and prevent another event similar to the 2015 GKM spill, as shown in Figure 9.1.

THE ADVANTAGES AND DISADVANTAGES OF ADVERSARIAL LEGALISM

Litigation is a central feature of American politics. It has played a crucial role in the struggle over civil rights, abortion, tobacco regulation, electoral redistricting, reforming criminal justice, making society more accessible to those with disabilities, and cleaning up the environment, to name some policy areas. Indeed, commenting on 1830s America, Alexis de Tocqueville (2003 [1835], 315) observed, "There is hardly a political question in the United States which does not sooner or later turn into a judicial one." Contemporary political science has expanded on this idea, supporting de Tocqueville's observations empirically as well as considering the consequences of this peculiar American phenomenon (Barnes and Burke 2012; Burke 2002; Kagan 2001; Melnick 1983, 1994; Silverstein 2009).

According to many of these scholars, the United States has become a "litigation state" (Farhang 2010) in which "juridification" (Silverstein 2009), "litigious policies" (Burke 2002), "adversarial legalism" (Kagan 2001), and "legalized accountability" (Epp 2009) have proliferated as a way to enforce compliance with federal and state laws. Thus the use of private litigation to achieve public policy goals has been well documented.

Some maintain that a legalistic culture—typically dubbed "adversarial legalism"—encourages legal action over disputes better settled by other government actors (Kagan 2001; Melnick 1983), that social movements use and are affected by their efforts to create policy victories through law and courts (Epp 2010; McCann 1994), and that elected officials—in creating statutes—incentivize and affect litigation behavior (Burke 2002; Farhang 2010). Still others have centered more on the work of court actors themselves and their judicial power to create policy change (Feeley and Rubin 1998; Frymer 2003). These so-called private enforcement regimes—statutory regimes in which private plaintiffs bring enforcement actions—occur across a broad swath of policy areas (Farhang 2010). Still, they share in common private rights of action for individuals to enforce the law through courts,

FIGURE 9.1. Toxic plume flows through Durango, Colorado. *Courtesy*, EcoFight/ Bruce Gordon

often with provisions for attorney fees to be recouped by successful plaintiffs or allowances for damages in multiples of the actual harm caused. Recently, political scientists have taken interest in these regimes, arguing that they are important examples of state power as expressed through the legal state (Frymer 2003).

The use of adversarial legalism falls into two categories, those that see it as positive and those that see it as detrimental to governance. Some scholars, most prominently Gerald N. Rosenberg (2008), find that the over-reliance on law and courts to enforce social policies diverts interest groups and social movements with limited time and resources away from more consequential and legitimate modes of political advocacy, such as grassroots mobilizing and lobbying.[3] Moreover, the use of litigation and the pursuit of legal rights take broad political grievances and political rights and place them in narrower, legalistic categories (Silverstein 2009, 69). In contrast, others, like Jeb Barnes and Thomas F. Burke (2015), maintain that the charges against adversarial legalism do not withstand empirical scrutiny. They consider four criticisms of adversarial legalism: "(1) that it prevents other forms of political action; (2) it creates path-dependent development, which forces government into bad policymaking in the future; (3) it produces 'polarizing backlashes'; and (4) it stymies 'social solidarity' by individualizing conflict" (Barnes and Burke 2015, 15). Barnes and Burke only find evidence for the fourth of these charges. Despite the debate, the evidence in the literature generally points to the positive aspects of adversarial legalism or, at the very least, that litigation has proven effective in reaching political outcomes, since elected officials are often glad to have political disputes between "disgruntled interests" settled elsewhere (Barnes and Burke 2015, 3; Burke 2002; Farhang 2010; Graber 1993; Kagan 2001; Lovell 2003).

Therefore, we certainly have a sense of how plaintiffs turn their claims into results, and we also have a great deal of knowledge about the consequences of adversarial legalism. Relatively absent in the conversation about adversarial legalism, however, is a discussion about *why* the system works this way, *how* we got here, and *which* areas of American society are best left to adversarial legalism.[4] More than that, when private actors become enforcers of US policy and wielders of state capacity, it is necessary to evaluate the use of that capacity. If the US Congress is using private litigation as a means of building state capacity, then studies of private enforcement regimes and citizen suits take on a central role in the story of American state building. Scholars in the American Political Development tradition have also recently noted the "porous boundaries" between state and non-state actors throughout the Progressive Era of state building (Balogh 2009; Nackenoff 2014, 132). We have continued to see non-state actors, such as citizens, advance the central state's capacity through these citizen suits into the twenty-first century.

Yet given that these suits and adversarial legalism "individualize interests" (Barnes and Burke 2015, 15), citizen enforcement of environmental standards is not the best way to solve these national problems. Congress's use of non-state actors to expand the American state has helped contribute to the kind of environmental blight seen with the Gold King Mine spill. In the first place, the General Mining Law of 1872, an attempt to expand the United States westward and to settle territory, gave rise to the ubiquity of mining out West; these abandoned mines are now endangering citizens and ecosystems of the West. In an attempt to remedy some of these environmental problems, more than a century later the US Congress passed the Clean Water and Superfund

Acts, which again gave citizens power to help expand and wield central state capacity. Ultimately, in the context of environmental regulation, citizen suits are not an effective way to remedy problems that should be recognized as nationwide, like the status of abandoned mines in the West.

HOW CITIZEN LITIGATION WORKS

Through litigation, citizens play an important role in enforcing the nation's environmental laws. Sixteen of the chief US federal environmental laws detail provisions by which citizens can sue as "private attorney general" to ensure compliance or to hold government agencies to performing their mandated duties (May 2004, 53). These so-called federal citizen suits allow "any person" to "commence a civil action on his own behalf" against either "any person" who violates the law's provisions or against the EPA for failing "to perform any act or duty . . . which is not discretionary" (May 2004, 53).[5]

Congress first created the citizen suit provision in its 1970 version of the Clean Air Act and then placed a similar provision in the Clean Water Act in 1972. Since then, it has reauthorized citizen enforcement in nearly every major piece of federal environmental legislation (Abell 1995, 1959). The federal courts have come to see citizen suits as "a deliberate choice by Congress to widen citizen access to the courts as a supplemental and effective assurance that [environmental laws] would be implemented and enforced."[6] Congress, too, envisioned that citizens would play a key role in enforcement (Abell 1995, 1960).[7] Studies have shown that citizens have come to play a "surrogate enforcement role" rather than the supplementary role envisioned by Congress. Thus with the rise of private enforcement, the Clean Water Act has become the most "popular tool" of citizen action (Abell 1995, 1960).[8] Between 1973 and 2002, citizens accounted for more than 1,500 reported federal decisions in civil environmental cases. And from 1993 to 2002, federal courts issued opinions, on average, on 110 civil environmental cases per year. Of these 110 cases, 83 were citizen suits, roughly 75 percent (May 2004, 54).

Despite the recent frequency of the citizen suits, private enforcement of laws has a long history. Until the nineteenth century, in England the majority of criminal prosecutions were initiated by private citizens because public prosecutors and police forces did not exist (Landes and Posner 1975, 2). While the United States does not have as strong a tradition of private enforcement of criminal statutes, these kinds of statutes have existed in this country ever since the First Congress passed six such laws. Private enforcement continues to play an important role in criminal statutes such as the False Claims Act, which dates back to the Civil War period and is the federal government's primary litigation tool to combat fraud against the government (Abell 1995, 1961). Private enforcement of laws typically gave citizens a personal financial stake in the judicial outcome, but citizen suit provisions in environmental law do not have comparable personal compensation for successful prosecution. Instead, in environmental law, those suits serve as a public good: when plaintiffs seek civil penalties under the Clean Water Act, those penalties are paid solely to the US Treasury.

Citizen enforcement of the Clean Water Act has become popular because it does not impose any restriction on the types of violations for which citizen plaintiffs may sue. And the limitations on citizen that it *does* pose are not much of an obstacle to citizen litigation. The Clean Water Act details two limitations. First, before suing, a citizen must file a note of intent to sue that details the alleged violation. Second, a citizen cannot sue if the EPA or the state has begun "diligently prosecuting" a civil or criminal action in court ("Navigation and Navigable Waters" 2011). In practice, however, these two limitations are "essentially toothless," leaving "citizen plaintiffs basically unchecked to exercise executive, prosecutorial authority as a 'private attorney general'" (Abell 1995, 1963–1964). The question remains, though: Why have citizen suits proliferated over time? That is, how have courts become so involved in environmental protection? The answer to this question is connected to the history and centrality of courts in environmental protection.

COURT INVOLVEMENT IN ENVIRONMENTAL PROTECTION

Judicial opinions often provide the foundation on which we discuss natural resource management and protection today. A sampling of federal judicial rulings demonstrates the importance of court-enforced environmental regulation: government can regulate the use of private property within certain parameters; citizens are often able to legally challenge agency actions that may harm the environment; agencies need to seriously consider reasonable alternatives to proposed actions when told to do so; the 1970 National

Environmental Policy Act (NEPA) is not a "paper tiger," and its procedural obligations are to be taken seriously; and the "plain intent" of Congress in enacting the Endangered Species Act (ESA) in 1973 "was to halt and reverse the trend toward species extinction, whatever the cost" (Nie 2008, 141).[9] The judiciary's role in environmental regulation in the United States has become so pronounced because, as Shep R. Melnick (2004, 103) concludes, "we distrust centralized bureaucracy, so we rely heavily on state and local agencies. But we expect these agencies to respect the basic rights of Americans and to meet minimum national standards. So we subject these agencies to substantial regulation by federal agencies. Because we do not trust these federal agencies, either, we demand judicial oversight."

Courts have become a central player in environmental oversight largely because of the Administrative Procedures Act (APA) of 1946. Suffice it to say that the APA was a compromise between President Franklin D. Roosevelt's administration, which wanted to give bureaucratic agencies vast rule-making power subjected to little judicial review, and Republicans and many in the American Bar Association, who wanted agencies to engage in the more time-consuming case-by-case adjudication subjected to strong judicial review (Melnick 2004, 91). The APA created two forms of agency action: "notice and comment rule making," in which agencies act like a legislature promulgating general rules; and "formal adjudication," where agencies act more like a judicial body deciding on a particular case. Courts must approve regulations created by notice and commenting rule making unless they are "arbitrary, capricious, an abuse of discretion, or otherwise not in accordance with law" ("Government Organization and Employees" 2011). In this case, the APA authorizes the judicial review of agency actions. While deference to agencies is most common, courts still examine the quality of an agency's reasoning. Whether courts use the more probing "hard look" standard or the more corporate-deferential "Chevron" inquiry,[10] they attempt to ensure that agencies' decisions are based on an administrative precedent and are sensible (Nie 2008, 141).[11]

More than that, as noted above, many environmental laws have citizen suit provisions, enabling interested parties to sue agencies and private companies believed to be violating the law. These provisions and the APA are designed to supplement government enforcement of environmental laws. Most of these provisions are patterned after section 304(a)(2) of the Clean Air Act, authorizing "any person" to sue the administrator of the EPA "where there is alleged a failure of the Administrator to perform any act or duty under this chapter which is not discretionary with the Administrator" ("Public Health and Welfare" 2009).

An expansive view of the rule of standing also helps explain the centrality of the judiciary in environmental litigation. Indeed, a growing number of parties have gained wider access to the courts following historic decisions. For example, *Scenic Hudson Preservation Conference v. Federal Power Commission* (354 F. 2d 608 [2d Cir. 1965]) granted environmental organizations standing to challenge a license to construct an electric-generating system on Storm King Mountain in New York's upper Hudson Valley. Similarly, *Sierra Club v. Morton* (405 U.S. 727, 1972) held that standing goes beyond economic harm and that it could be granted if environmental interests showed aesthetic or ecological harm. Consequently, the judiciary has become quite receptive to environmental claims (Nie 2008, 142).

Statutory ambiguity, too, requires the courts to play a central role in enforcement and compliance (*Scenic Hudson* 1965). While some laws "breathe discretion at every pore," many also contain standards and binding obligations that are judicially enforceable.[12] Such ambiguity contributes to an American culture of "adversarial legalism." As shown by Robert A. Kagan (2001), the US model of policymaking, implementation, and dispute resolution is characterized by frequent appeals to adversarial legal contestation. This is due to a number of factors, including the nature of American laws that are comparatively complex, vague, and indeterminate. Kagan (2004, 26–27) also shows how American political culture demanded a more active government in the 1970s, "one that would enforce nationwide standards of justice and of security from harm." Americans, however, still maintained their distrust of centralized political authority, which explains the ubiquity of court-based enforcement in the United States.

Many scholars note that litigation alone is "politically inadequate" (Nie 2008, 143). As natural resource scholar Martin Nie (2008, 143) concludes, "It often serves as a shield, or defense mechanism, rather than a sword, or offensive weapon." Thus more sweeping legislative changes, rather than the much slower and individualized judicial process, could offer better solutions. Before discussing solutions, though, we must examine some of the problems that confront Western states seeking to remedy the legacy of hardrock mining.

LEGACY OF HARDROCK MINING AND ITS PROBLEMS

The western United States experienced extraordinary mineral development in the late nineteenth and early twentieth centuries, and for most of this period there were few expectations about environmental safeguards. Ginny Brannon, director of the Colorado Division of Reclamation, Mining, and Safety, noted that Colorado had minimal environmental regulations governing mining until 1977 (quoted in "San Juan County Spill" 2015). In 1891, the federal government only had two restrictions on opening a mine: provide adequate ventilation and do not employ people under the age of twelve (*History of Mine Safety* n.d.). That was it. There were no inspections and certainly no environmental standards. State regulations were equally sparse. Pollution simply wasn't a major concern in the era when Gold King—and thousands of other now abandoned mines—operated. The 1872 General Mining Law, which still governs hardrock mining, sought to encourage westward expansion but created an array of "unintended consequences of unrestrained environmental pollution" (Bakken 2008, 36).

But with the heightened consciousness of the environmental costs of mining in the 1960s, the EPA began to regulate mining in the 1970s as part of the Clean Water Act. Modern mines are therefore required to plan for treating any water degraded by mining. The Clean Water Act, as detailed above, also made provisions for citizen suits to enforce the standards of the federal law. However, abandoned mines are sites where minerals or ores were extracted prior to these Clean Water regulations, and, more important, these mines likely do not have owners who can be held responsible for the mining activities that occurred there. When Colorado—the site of the Gold King Mine spill—last prepared a list of priority abandoned mine restoration projects in 2012, it included only seven mines. The cost to repair and maintain each of those mines ranged from $50,000 to upward of $1 million, based on figures from the Water Quality Control Division of the Colorado Department of Public Health and the Environment ("Gold King Mine" 2015).

Yet abandoned mines are spread across Colorado. The state has a branch within its Department of Natural Resources called the Inactive and Abandoned Mine Reclamation Program. The head of the branch, Bruce Stover, said the state has been working for years to address the issue of pollution from approximately 22,000 abandoned mines. Stover said the branch tries to "go in and characterize which mines are the worst offenders. Is it this drain over here, is it that waste pile over there? And we try to do projects that incrementally chip away at the overall problem." With this method, Stover claimed it would take "decades" to clean up just three-quarters of the mines leaking hazardous materials (quoted in "Animas River Spill" 2015).

Part of the problem is that there is not nearly enough money to pay for cleanups. Colorado estimates that it spends $5,000 to deal with each problem area in a mine. Since the state has 10,818 locations that pose safety hazards, such as open mine shafts, in addition to its 4,670 environmental threats, addressing these issues will eventually cost around $80 million—and the state's annual budget for cleanup is $2 million ("Hey, Congress" 2015). Stover notes that without owners, there is no responsible party to pay: "It's a huge problem in Colorado because these are old, abandoned active mines and they don't have any owners and they are just draining." And so we are left wondering: Who pays for cleaning up mines without owners? Funding for cleanup often comes from the EPA and other federal sources. The program that administers water reclamation projects, Colorado's Nonpoint Source program, planned to use $1 million from the EPA for fiscal year 2015 "for implementation projects that restore impaired waters" (Stover quoted in "Animas River Spill" 2015). And the Colorado Division of Reclamation, Mining, and Safety gets about $2 million for mine safeguarding work from the 1977 Surface Mining Control and Reclamation Act ("Gold King Mine" 2015). But as Stover said, "The state doesn't really have the money to tackle these draining mines" (quoted in "Animas River Spill" 2015).

Notwithstanding the dearth of cleanup funds, the chief problem is that under existing federal laws, liability is *strict*, *joint and several*, and *retroactive* (Tilton 1994, 64–65). Liability is *strict*, meaning that a defendant (either a miner owner of the land or the party involved in the mine cleanup) is in legal jeopardy by virtue of the existence of a wrongful act, without any accompanying intent or mental state. That is, whether a defendant intended to cause damage to another's property or to a waterway is irrelevant, and the defendant will be held accountable. Federal law governing water pollution is also *joint and several*, meaning that any single potentially responsible party can be held liable for the entire cost of the remediation effort. Finally, liability is *retroactive*, meaning that cleanup standards devised in the 1980 Superfund legislation, for example, apply retroactively to the generation, transport, and storage of wastes that occurred before 1980.[13]

These laws concerning liability govern the primary piece of congressional legislation used to clean up hardrock mines: the Superfund law. Many hardrock mining sites are currently abandoned or have been inactive for long periods and thus cannot be easily regulated under public mining laws or pollution control laws aimed at current operating facilities (Seymour 2004, 800). To remedy this problem, federal regulators often use their authority under the Superfund law. With its strict, retroactive, and joint and several liability scheme, the Superfund statutory authority allows the United States to recover its costs from broadly defined categories of the parties responsible—without regard to their negligence as to whether the activities that gave rise to the contaminant were lawful or even consistent with best management practices existing at that time (Seymour 2004, 802). Using Superfund authority, the EPA has added some of the nation's largest and most seriously contaminated mining sites to the National Priorities List (NPL), a list of sites schedule for long-term remedial action.[14] Indeed, in the wake of the Gold King Mine spill, Silverton and San Juan County—the site of the disaster's onset—voted yes for a Superfund cleanup of old mines, reversing decades of opposition to such a designation and thus demonstrating that the severity of the Gold King Mine spill and the broader environmental problem had polluted waterways in the West ("Silverton, San Juan County Vote Yes" 2016). But serious problems remain.

Superfund enforcement at mining sites on public lands in particular presents immense challenges. Most of the nation's hardrock minerals are found in twelve Western states, and many hardrock mines are located, at least in part, on public lands (Seymour 2004, 804). While the EPA is the primary authority for CERCLA cleanups on private lands, former US president Donald Trump delegated the lion's share of Superfund authority on public lands to federal land managers with jurisdiction, custody, and control over those lands. Superfund status waives the federal government's sovereign immunity to lawsuits and thus makes federal agencies attempting cleanup on public land liable in the same manner and to the same extent as any nongovernmental entity. Thus federal and private entities alike can be found liable for cleanup costs under CERCLA if they are "owners," "operators," or parties who "arrange for disposal of these contaminated substances at a facility" (Seymour 2004, 805, quoting from the CERCLA statute).

The complex web of policy and legal structures makes CERCLA enforcement difficult. When a federal agency exercises its delegated CERCLA authority, mining companies often allege that such "jurisdiction, custody, or control" is enough to suggest that the United States should be liable as a site "owner." In addition, when the EPA attempts to enforce action on hardrock mines on commingled private/public lands, mining owners often file third-party actions against the United States, arguing that federal land managers are liable as co-owners of a mining "facility" or as "operators" at the facility. Mining companies have claimed that the federal government's regulation of private mining companies and its involvement in encouraging wartime production minerals make federal agencies liable as "operators" of mining facilities or as parties that "arrange for" the disposal of hazardous material at the facility (Seymour 2004, 805). Indeed, this fight over liability is taking place currently with the Gold King Mine; its owner, Todd Hennis, argues that he is not responsible for the spill, while the EPA is attempting to make him a "potentially responsible party" (quoted in "Public Bill" 2015). Ultimately, because economic penalties and liability are both severe and strict, we are left with protracted legal battles and cleanup impasses.

All of these provisions tend to deter cleanups because, as economist David Gerard (Property and Environment Research Center 2015) notes, "One set of rules is used to address two distinct tasks—the remediation of past pollution and deterrence against future pollution." This means that groups that initiate cleanup efforts are treated no differently, under both the Superfund and Clean Water Acts, than the groups that caused the pollution in the first place. Therefore, volunteer parties can be held liable for the extent of all abandoned mine cleanup if they take on remediation efforts. Mine cleanup often requires discharges that can pollute rivers and lakes, discharges the EPA must permit as detailed in the Clean Water Act. Parties that affect the discharge must be permitted; through that process, these parties assume legal responsibility for meeting the EPA permit standard. As Gerard observes, "The assignment of liability occurs even if the remediating party had no role in generating the pollution, and even if the party had nothing to do with generating the water-quality degradation" (Property and Environment Research Center 2015). Given the difficulty of mine cleanup, as we saw with the Gold King Mine spill caused by the EPA itself, it comes as no surprise that few environmental groups and private parties are rushing in to clean up the widespread abandoned mines throughout the United States.

POTENTIAL SOLUTIONS

If anything was made clear by the Gold King Mine spill, it was that the laws governing hardrock mining cleanup cannot respond to the vast economic and environmental liabilities confronted by mountain communities throughout the West. The regulatory environment that governs how mining occurs—and the legal liability assigned to various parties for reclamation, damage, and cleanup—disincentivizes resolving these environmental problems. The General Mining Law of 1872 does not mandate a firm legal requirement or enough money to clean up these old mines. In addition, when we add to this the financial liability that would be assigned should any cleanup go wrong, we get our current situation: a system ill-equipped to meaningfully enforce federal laws governing acid mine drainage. Fixing the problem requires a multifaceted legislative approach, an approach that would help create the necessary funds as well as mitigate liability for those seeking to clean up mines.

The California Gold Rush and other Western mining booms of the mid-nineteenth century helped create the 1872 General Mining Law. In the West, mineral deposits were found predominantly on federal lands, but there existed no law governing the transfer of these mineral rights from public to private owners. Thus in 1872, Congress codified the customs, codes, and laws miners had been using in previous decades. This congressional legislation gave broad discretion over the use of public land resources to the private companies, requiring very little government oversight. The central provisions of this legislation, remarkably, remain intact today (Gerard 1997; Seymour 2004, 825).[15]

The General Mining Law allows US citizens and firms to explore for minerals and establish rights to federal lands without authorization from any government agency. If a site contains a deposit that can be profitably marketed, claimants enjoy the "right to mine," regardless of other potential uses or non-use value of the land (Gerard 1997). Originally, claimants maintained their right to mine by satisfying an annual work requirement, but in 1992 the US Congress replaced this requirement with an annual $100 holding fee for each claim. With this fee, claimants acquire outright title to both the minerals and the land by obtaining a mineral patent at a per-acre cost of $2.50-$5.00. Unlike natural gas or coal mining, where operators must pay for the right to extract minerals, no such rules govern hardrock mining; producers do not pay royalty taxes on the minerals taken from federal lands.

Given the lack of cleanup funding for hardrock sites (Lounsbury 2008; Seymour 2004), a hardrock reclamation fund might be one part of a long-term solution. A portion of the proposed royalties, rents, or taxes paid by hardrock miners could supply a fund dedicated to the cleanup of inactive and abandoned mines. Congress could follow the legislation it created to clean up abandoned coal mines. The Abandoned Mine Reclamation Fund has gathered its money from a per-ton tax on coal mined in the United States, collecting $7.4 billion from January 1978 to October 2005 (Lounsbury 2008, 199). The Hardrock Mining and Reclamation Act, introduced November 5, 2015, is a good start in solving the issues of funding. Among other things, the bill would set a 2 percent–5 percent royalty fee on new hardrock mining operations and establish a hardrock minerals reclamation fund to help finance abandoned mine cleanup.[16]

In terms of liability, the disincentive to reclaim old sites extends not only to private parties but also to state governments. If a state begins cleanup of an old site, it is required to reduce pollution levels to those specified by the Clean Water Act, regardless of cost. Faced with this level of cleanup or nothing at all, states often have an incentive to do nothing. As noted above, under CERCLA, liability is joint and several, strict, and retroactive; it extends to parties classified as current owners or operators, owners or operators at the time of disposal, generators, arrangers, and transporters. As a result, a Good Samaritan who, for example, removes a small amount of acidic drainage leaching into a river and caps it elsewhere might become liable for remediating the *entire* site, including all hazardous residue generated by historical mining operations. The economic costs of such liability are staggering. CERCLA cleanups even at "non-mega"-mine sites usually cost in the seven figures (one Congressional Research Service study estimated the average cost to be $22 million), and reclamation of "mega"-sites can cost hundreds of millions of dollars (Lounsbury 2008, 153). Given the potential for this sort of liability, it is understandable that Good Samaritans would refrain from action to avoid the repercussions of CERCLA. Ironically, these strict environmental regulations have had the unintended consequences of deterring cleanup.[17] Thus the shortcomings of environmental liability laws should be at the center of any reform debate, but they are not.

Thus Representative Scott Tipton (R-Cortez) is right to note the importance of Good Samaritan legislation, which would protect those who take on mine cleanup from the liability associated with the effort or the particular mine's problems in the first place. Senator Michael Bennet (D-CO) and former senator Mark Udall introduced a bill that would protect third-party cleanup groups from liability under the Clean Water Act, but that measure did not pass the US Senate. Tipton had sponsored companion legislation in the US House. Former senator Cory Gardner later took up the cause along with Bennet and Tipton, a cause that seems the most useful in solving abandoned mine issues ("Colorado Sens. Bennet, Gardner Urge Law" 2016).

In speaking in support of Good Samaritan legislation, Senator Bennet said recently, "The Gold King Mine spill has served [as] a catalyst to focus Congress' attention on the dangers posed by the thousands of abandoned mines in Colorado and throughout the West" ("Colorado Sens. Bennet, Gardner Urge Law" 2016). Other senators, like Barbara Boxer (D-CA), have expressed support for this solution but have noted "problems" with the proposed solution. Boxer said, "I want to make improvements to the legislation so that it will protect the environment and ensure that taxpayers will not be on the hook if a good Samaritan makes the pollution" ("Colorado Sens. Bennet, Gardner Urge Law" 2016). Certainly, the resounding claim from environmental, legal, and political science scholars has supported Bennet's urge for change and the concerns Boxer has expressed. Let's hope Congress can deliver this time around.

NOTES

1. Proposed by Senators Michael Bennet (D-CO) and Cory Gardner (R-CO) and Representative Scott Tipton (R-CO) in 2016.

2. Introduced by Senator Tom Udall (D-NM) in 2015.

3. Some groups, however, successfully pair their legal work with other grassroots organizing in multifaceted campaigns in which the various elements reinforce each other. For example, the Center for Biological Diversity has made litigation using the Freedom of Information Act (FOIA, 1967), the Administrative Procedures Act (1946), and especially the Endangered Species Act (1973) central to its strategy and identity. Indeed, a legal threat can serve as ammunition for legislative change and spur turnout at legislative, rule-making, and other hearings. Pending or active lawsuits can attract new members and cultivate leaders and a fundraising base that supports the legal strategy. See Ferguson and Hirt (2019).

4. Shep R. Melnick's (1999) work is a good exception, as he demonstrates the dangers of adversarial legalism in tobacco regulation.

5. May (2004) quotes from both the Clean Air Act, 42 U.S.C. §7604(a) (2000), and the Clean Water Act, 33 U.S.C. §1365(a) (2010).

6. *Natural Resources Defense Council v. Train*, 510 F.2d. 692, 700 (D.C. Cir. 1974). Citizen suits do not apply only to federal enforcement. A 1997 survey found that twenty-six states allow citizens to enforce state environmental laws (May 2004, 53).

7. During a 1973 congressional debate over revisions to the Clean Water Act, Senator Birch Bayh (D-IN) said that citizen suits "can be a very important tool for keeping industry and Government alike from letting standards and enforcement slip" (quoted in Abell 1995, 1960n11). In an interview, Kieran Suckling, director of the Center for Biological Diversity, said that the citizen participation and appeal provisions of various environmental laws are the most valuable tools his group has for protecting endangered species and habitat. He also said that probably the most important law they use with the greatest frequency is FOIA. See Ferguson and Hirt (2019).

8. Early studies of the Clean Water Act show that "more than 65 percent of all citizens actions . . . had been brought under the Clean Water Act" (Miller 1987, quoted in Abell 1995, 1960n13).

9. Martin Nie (2008, 141) notes these judicial opinions. In corresponding order, see *Penn Central Transportation Co. v. New York City*, 438 U.S. 104 (1978) (takings); *Scenic Hudson Preservation Conference v. Federal Power Commission*, 354 F. 2d 608 (2d Cir. 1965) (standing); *Citizens to Preserve Overton Park v. Volpe*, 401 U.S. 402 (1971) (consideration of alternatives); *Calvert Cliffs' Coordinating Committee v. United States Atomic Energy Commission*, 449 F.2d 1 109 (D.C. Cir. 1971) (NEPA); and *Tennessee Valley Authority v. Hill*, 437 U.S. 153, 184 (1978) (ESA).

10. Named for *Chevron U.S.A., Inc. v. Natural Resources Defense Council, Inc.*, 467 U.S. 837 (1984), in which the US Supreme Court articulated a set of legal tests to determine whether to grant deference to a government agency's interpretation of a congressional statute.

11. For judicial decisions that define these standards of review, see *Greater Boston Television Corp. v. FCC*, 444 F. 2d 841, 850–852 (D.C. Cir. 1970), cert. denied, 403 U.S. 923 (1971); *Motor Vehicle Manufactures Association v. State Farm Auto Insurance Co.*, 463 U.S. 29, 43–44 (1983); *Chevron* (1984); and *Overton Park* (1971).

12. Nie (2008, 142), quoting from *Strickland v. Morton*, 519 F. 2d 467, 468 (9th Cir. 1975).

13. For more discussion of the impacts of these legal guidelines, see Property and Environment Research Center (2015).

14. See *Mining and Mineral Processing Sites* (1997). Nearly all of the more than sixty sites then listed on the NPL had, at least in part, been previously used for the

extraction, processing, or beneficiation of hardrock minerals. For a case study of this Superfund process, see Brooks (2015).

15. For an excellent history of the 1872 General Mining Law, see Wilkinson (1993), chapter 2. An even more sustained analysis can be found in Bakken (2008).

16. Introduced by Senator Tom Udall (D-NM). Unfortunately, its last action was on November 5, 2015: it was read twice and referred to the Committee on Energy and Natural Resources.

17. For a discussion of liability and its politics, see Brooks (2015), chapter 3.

REFERENCES

Abell, Charles S. 1995. "Ignoring the Trees for the Forests: How the Citizen Suit Provision of the Clean Water Act Violates the Constitution's Separation of Powers Principle." *Virginia Law Review* 81 (7): 1957–1987.

"Animas River Spill: A Stark Reminder of Colorado's Mine Pollution Legacy." 2015. Community Radio for Northern Colorado, August 11. http://www.kunc.org/post/animas-river-spill-stark-reminder-colorados-mine-pollution-legacy#stream/0.

Bakken, Gordon. 2008. *The Mining Law of 1872: Past, Politics, and Present*. Albuquerque: University of New Mexico Press.

Balogh, Brian. 2009. *A Government Out of Sight: The Mystery of National Authority in Nineteenth-Century America*. New York: Cambridge University Press.

Barnes, Jeb, and Thomas F. Burke. 2012. "Making Way: Legal Mobilization, Organizational Response, and Wheelchair Access." *Law and Society Review* 46 (1): 493–524.

Barnes, Jeb, and Thomas F. Burke. 2015. *How Policy Shapes Politics: Rights, Courts, Litigation, and the Struggle over Injury Compensation*. New York: Oxford University Press.

Brooks, David. 2015. *Restoring the Shining Waters: Superfund Success at Milltown, Montana*. Norman: University of Oklahoma Press.

Burke, Thomas F. 2002. *Lawyers, Lawsuits, and Legal Rights: The Battle over Litigation in American Society*. Berkeley: University of California Press.

"Colorado Sens. Bennet, Gardner Urge Law to Spur Cleanup at Old Mines." 2016. *Denver Post*, March 2. http://www.denverpost.com/news/ci_29587008/colorado-sens-bennet-gardner-urge-law-spur-cleanup?source=pkg.

Epp, Charles R. 2009. *Making Rights Real: Activists, Bureaucrats, and Supreme Court in Comparative Perspective*. Chicago: University of Chicago Press.

Epp, Charles R. 2010. "Law's Allure and the Power of Path-Dependent Legal Ideas." *Law and Social Inquiry* 35 (4): 1041–1052.

Farhang, Sean. 2010. *The Litigation State: Public Regulation and Private Lawsuits in the United States*. Princeton, NJ: Princeton University Press.

Feeley, Malcolm M., and Edward Rubin. 1998. *Judicial Policy Making and the Modern State: How the Courts Reformed America's Prisons*. New York: Cambridge University Press.

Ferguson, Cody, and Paul Hirt. 2019. "'Power to the People': Grassroots Advocacy for Environmental Protection and Democratic Governance in the Late Twentieth Century." In *The Nature of Hope*, edited by Char Miller and Jeffrey Crane, 52–78. Boulder: University Press of Colorado.

Frymer, Paul. 2003. "Acting When Elected Officials Won't: Federal Courts and Civil Rights Enforcement in US Labor Relations, 1935–1985." *American Political Science Review* 97 (3): 483–499.

Gerard, David. 1997. *The Mining Law of 1872: Digging a Little Deeper*. Policy Series, PS 11. Bozeman, MT: Property and Environment Research Center.

"The Gold King Mine: From an 1887 Claim, Private Profits, and Social Costs." 2015. Colorado Public Radio, August 17. http://www.cpr.org/news/story/gold-king-mine-1887-claim-private-profits-and-social-costs.

"Gold King Mine Spill: Six Months Later." 2016. *Durango Herald*, February 7, 1A.

"Government Organization and Employees." 2011. Title 5 U.S.C., §706, Supp. V, 112. https://www.gpo.gov/fdsys/granule/USCODE-2011-title5/USCODE-2011-title5-partI-chap7-sec706.

Graber, Mark. 1993. "The Non-Majoritarian Difficulty: Legislative Deference to the Judiciary." *Studies in American Political Development* 7: 35–73.

"Hey, Congress, Quit Blaming the EPA: The Gold King Mine Accident Is on You, Too." 2015. *On Earth*, September. http://www.onearth.org/earthwire/congress-blaming-epa-for-animas-river-spill.

History of Mine Safety and Health Legislation. n.d. US Department of Labor. http://arlweb.msha.gov/MSHAINFO/MSHAINF2.htm.

Kagan, Robert A. 2001. *Adversarial Legalism: The American Way of Law*. Cambridge, MA: Harvard University Press.

Kagan, Robert A. 2004. "American Courts and the Policy Dialogue: The Role of Adversarial Legalism." In *Making Policy, Making Law: An Interbranch Perspective*, edited by Mark C. Miller and Jeb Barnes, 13–34. Washington, DC: Georgetown University Press.

Landes, William M., and Richard A. Posner. 1975. "The Private Enforcement of Law." *Legal Studies* 4 (1): 1–46.

Lounsbury, Bart. 2008. "Digging Out of the Holes We've Made: Hardrock Mining, Good Samaritans, and the Need for Comprehensive Reform." *Harvard Environmental Law Review* 32: 149–216.

Lovell, George. 2003. *Legislative Deferrals: Statutory Ambiguity, Judicial Powers, and American Democracy*. New York: Cambridge University Press.

May, James R. 2004. "The Availability of State Environmental Citizen Suits." *Natural Resources and Environment* 8 (4): 53–56.

McCann, Michael. 1994. *Rights at Work: Pay Equity Reform and the Politics of Legal Mobilization*. Chicago: University of Chicago Press.
Melnick, Shep R. 1983. *Regulation and the Courts: The Case of the Clean Air Act*. Washington, DC: Brookings Institute Press.
Melnick, Shep R. 1994. *Between the Lines: Interpreting Welfare Rights*. Washington, DC: Brookings Institute Press.
Melnick, Shep R. 1999. "Tobacco Litigation: Good for the Body But Not the Body Politic." *Journal of Health Politics, Policy, and Law* 24 (4): 805–810.
Melnick, Shep R. 2004. "Courts and Agencies." In *Making Policy, Making Law: An Interbranch Perspective*, edited by Mark C. Miller and Jeb Barnes, 89–104. Washington, DC: Georgetown University Press.
Miller, Jeffrey G. 1987. *Citizen Suits: Private Enforcement of Federal Pollution Control Laws*. New York: Wiley Law Publications.
Mining and Mineral Processing Sites on the NPL. 1997. Environmental Protection Agency. https://www3.epa.gov/epawaste/hazard/tsd/ldr/mine/npl.pdf.
Nackenoff, Carol. 2014. "The Private Roots of American Political Development: The Immigrants' Protective League's 'Friendly and Sympathetic Touch,' 1908–1924." *Studies in American Political Development* 28 (2): 129–160.
"Navigation and Navigable Waters." 2011. Title 33 U.S.C., §1319, Supp. V, 418. https://www.gpo.gov/fdsys/granule/USCODE-2011-title33/USCODE-2011-title33-chap26-subchapIII-sec1319/content-detail.html.
Nie, Martin. 2008. "The Underappreciated Role of Regulatory Enforcement in Natural Resource Conservation." *Policy Sciences* 41 (2): 139–164.
Property and Environment Research Center. 2015. Interview with economist David Gerard, August 13. http://www.perc.org/blog/why-its-so-hard-clean-abandoned-mines.
"Public Bill for Gold King Spill Looms, EPA Seeks Liable Owners." 2015. *Denver Post*, December 25. http://www.denverpost.com/news/ci_29309457/public-bill-gold-king-spill-looms-epa-seeks.
"The Public Health and Welfare." 2009. Title 42 U.S.C., §7604, Supp. III, 6018. https://www.gpo.gov/fdsys/granule/USCODE-2009-title42/USCODE-2009-title42-chap85-subchapIII-sec7604.
Rosenberg, Gerald N. 2008. *The Hollow Hope: Can Courts Bring about Social Change?* 2nd ed. Chicago: University of Chicago Press.
"San Juan County Spill Highlights Years of Colorado Cleanup Effort." 2015. *Denver Post*, August 6. http://www.denverpost.com/environment/ci_28597473/san-juan-county-spill-highlights-years-colorado-cleanup.
Seymour, John F. 2004. "Hardrock Mining and the Environment: Issues of Federal Enforcement and Liability." *Ecology Law Quarterly* 31 (4): 795–956.
Silverstein, Gordon. 2009. *Law's Allure: How Law Shapes, Constrains, Saves, and Kills Politics*. Cambridge: Cambridge University Press.

"Silverton, San Juan County Vote Yes for Superfund Cleanup of Old Mines." 2016. *Denver Post*, February 22. http://www.denverpost.com/news/ci_29547638/silverton-san-juan-county-vote-possible-superfund-cleanup.

Tilton, John E. 1994. "Mining Waste and the Polluter-Pays Principle in the United States." In *Mining and the Environment: International Perspectives on Public Policy*, edited by Roderick G. Eggert, 57–84. Washington, DC: Resources for the Future.

Tocqueville, Alexis de. 2003 [1835]. *Democracy in America*. New York: Penguin Books.

Tuhus-Dubrow, Rebecca. 2015. "Climate Change on Trial." *Dissent* 62 (4): 152–158.

Wilkinson, Charles F. 1993. *Crossing the Next Meridian: Land, Water, and the Future of the West*. Washington, DC: Island.

10

Divergent Perspectives on AMD Remediation in the Upper and Lower Animas Watersheds

Pre- and Post-Spill Policy Preferences

BRAD T. CLARK

This chapter examines the two primary perspectives regarding the problem of acid mine drainage (AMD) in San Juan County, Colorado, and how to most effectively address water quality and biological conditions in the Animas River—both prior to and in the aftermath of the 2015 Gold King Mine (GKM) spill. In many respects, this involves a tale of two watersheds—those of the Upper and Lower Animas River—and how normative perspectives on AMD remediation have largely been divided between the two. In the Upper watershed, there has been an enduring preference to avoid Superfund designation, coupled with opposition to the strong presence of the federal government that would follow. This has been particularly evident among Silverton residents, who have expressed a preference for local- and state-led restoration—especially in the pre-spill context. A primary outlet for this position has been the Animas River Stakeholders Group (ARSG), a regional watershed-based group formed in 1994 to improve water quality and aquatic habitats through locally driven, collaborative processes ("Mission" 2017).

DOI: 10.5876/9781646421756.c010

In the Lower watershed, there has been preference—albeit less pronounced—for comprehensive, federally led remediation through Superfund listing of abandoned mine sites in the Silverton, Colorado, area. This has been expressed by select Durango residents, particularly those with interests linked to agriculture, tourism, and water-based recreation on the Animas—especially in the post-spill context.

A broad research expectation behind this project was that two primary coalitions existed *prior* to the 2015 mine spill, each articulating and promoting a distinct policy preference for AMD remediation. In this chapter, the Advocacy Coalition Framework (ACF) is broadly utilized, first to determine the relative presence of two competing coalitions and, second, to examine their policy preferences, both prior to and following the GKM spill.

THE ADVOCACY COALITION FRAMEWORK

The ACF was developed in the early 1980s by political scientists Paul Sabatier (1988) and Hank Jenkins-Smith as a "comprehensive approach to understanding politics and policy change over time" by "shed[ding] light on ideological disagreement and policy conflict" between competing coalitions (Jenkins-Smith et al. 2018, 136). The basis of unity for actors in each coalition is a common belief system that grounds their policy preferences and informs related strategies. A broad assumption behind the ACF is that for cases involving potential policy change, there exists prima facie two primary coalitions with distinct policy preferences—one generally in favor of maintaining the status quo and the other generally interested in advancing changes to existing policy.

As a potential pathway to policy change, the ACF posits an external event or dramatic shock to the status quo, which may prompt increased public and political attention to a given situation. This exposure and increased scrutiny may enable a coalition that favors change to successfully overcome the inertia behind the status quo and ongoing policy stasis. The widespread public and political attention following the GKM spill revealed that the status quo of AMD in the Upper watershed was unacceptable; the ACF suggests this would bolster efforts of the coalition seeking Superfund designation and cause members of the competing coalition to reconsider their opposition. Taken together, the ACF presents such conditions as providing an impetus for substantive policy change in the aftermath of a dramatic focusing event (see chapter 3, this volume).

Components of the ACF

The broad, theoretical interest of the ACF is to explain periods of both policy stability and abrupt change. To understand the policy process in between these periods, the unit of analysis is the *policy subsystem*, which is defined by a number of components (Jenkins-Smith et al. 2018, 139).

THE POLICY SUBSYSTEM

The first component of the subsystem relates to territorial scope, including the physical characteristics of the area in which a policy debate occurs. As discussed in chapter 1, the GKM case involves two distinct locations. One is the Upper Animas watershed in San Juan County, where the legacies of hardrock mining include abandoned mines, waste piles, and pollution. This area is the primary source of the AMD problem. The second location is the Lower Animas watershed, where downriver water quality has been negatively impacted by heavy metals released from the Silverton area. This is the area with the highest regional population densities; it is also the closest in proximity to downriver Native American communities.

The subsystem's second component is a specific policy topic. In the GKM case, the area of interest is environmental regulation—specifically, remediation of AMD from abandoned hardrock mining areas and subsequent restoration of aquatic environments. Included in this policy topic is the range of outputs (i.e., restoration options or strategies) and related outcomes (i.e., goals or results achieved) preferred by various actors or coalitions of actors.

The third subsystem component involves the political and institutional characteristics unique to each location, as well as the attributes of actors involved in the policy debate—including their preferences and political resources. In the GKM case, this involves the local governments and elected officials of San Juan and La Plata Counties, their demographics, and the political cultures unique to each location (for reference, see chapter 1). A crucial aspect of this political and institutional component is a broad understanding of the sets of relevant actors.

BELIEF SYSTEMS

The belief systems behind coalitions in the ACF guide information processing among actors and inform their understandings of the world. In the process, actors often simplify and organize the myriad stimuli to which they are exposed. As a result, all actors are subjective and motivated by some degree of self-interest in pursuit of specific goals.

The ACF presumes that policy actors in a given subsystem have belief systems composed of three tiers, each with a different degree of susceptibility to change (Jenkins-Smith et al. 2018, 140–145). The first, innermost tier involves so-called *deep core beliefs*, which are based on normative views on human nature and political ideology. These views include the appropriate role and scope of government in society. Deep core beliefs are firmly held by individuals and are often unchangeable; they form the basis and unity for advocacy coalitions in the ACF.

So-called *policy core beliefs* constitute the second tier. They are most often policy-specific and are said to be "bound by scope and topic to the subsystem [thus] hav[ing] territorial and topical components" (Jenkins-Smith et al. 2018, 140). Examples may include individuals' policy priorities, assessments of problem severity, root causes of problems, and preferred or acceptable solutions. The degree of susceptibility to change in these beliefs is considered moderate.

Third, at the outermost tier are so-called *secondary beliefs*. Unlike the previous two, these beliefs are not as deeply held by actors and are said to inform individuals' preferred "means for achieving desired outcomes outlined in [their] policy core beliefs" (Jenkins-Smith et al. 2018, 141). Beliefs at this level are posited as the most susceptible to change following a focusing event, the release of new information, or both (Wieble et al. 2011).

Magnitudes of Policy Change

Not all policy changes are equal; rather, they come in magnitudes or degrees of change to existing policies and the status quo. Leading ACF scholars refer to this as the "directionality of policy evolution" and argue that there exist different levels of policy change—just as there are different levels or tiers of belief systems (Jenkins-Smith et al. 2018, 144). Since the ACF presumes that policies and government programs are manifestations of policy-oriented beliefs, a change in beliefs represents a change in policy (Jenkins-Smith et al. 2018, 144). Hence the distinction between major and minor policy change.

To reiterate, deep core beliefs held by coalitions are unlikely to change. Thus major policy change is generally less likely to occur than minor policy change (Sabatier 1988). Yet major changes—such as National Priorities List (NPL) Superfund listing of the Bonita Peak Mining District (BPMD)—remain a possibility, given a set of certain conditions or considerations.

The conditions associated with major policy change stem from the impacts to coalitions in a subsystem following a focusing event. For example, actors in a coalition may react to an event by reevaluating their core beliefs, perhaps realizing that existing policy (i.e., the status quo) has failed. This may lead some to adopt policy preferences of the rival coalition. If a majority of these actors effectively endorse the other coalition's secondary *and* policy core beliefs, the ACF considers the balance of power between coalitions to have shifted; the pro-change coalition's belief system and core policy preference are now seen by the majority of actors as the most effective (and perhaps only remaining) means of addressing the problem. The coalition previously in support of the status quo may no longer be able to defend existing policy; policy change occurs.

THE ACF: UPPER AND LOWER ANIMAS RIVER WATERSHEDS

As previously stated, the decades-long saga on how to remediate AMD and improve water quality involves two watersheds of the Animas River—the Upper and Lower—and their divergent policy preferences. In the Upper watershed, opposition to the strong federal presence that is part of the Superfund process has largely been the preference of those residing in or with vested interests in San Juan County. Their preference has been for a local, stakeholder-driven approach to restoration.

The Pro-Status Quo Coalition—Resist Federal Takeover and Superfund Designation

Prior to the GKM spill, a set of core beliefs consistent with the policy preference of this coalition would have included support for main tenets of free market capitalism (limited intervention, free enterprise, and private property rights); libertarianism (individual liberty, freedom of choice); limited government (less intrusion); states' rights; and new federalism (devolution of power from federal to state governments).[1] These beliefs capture basic aspects of a belief system that is consistent with a policy preference for locally driven (i.e., nonfederal), stakeholder-based, and collaborative approaches to AMD remediation.

It is difficult, perhaps impossible, to fully determine the degree of change in this coalition's system of beliefs in the current *post*-spill environment; the assumption is that many actors continue to espouse many of their deep core beliefs—in particular those behind an aversion to comprehensive, federally led restoration. Yet in the post-spill context, the seriousness of AMD in San Juan County is clear and indisputable. Following the dramatic events of August 5, 2015, actors in the Upper watershed coalition were forced to confront a new reality; the cat was out of the bag, so to speak. Momentum quickly grew for significant policy change; it was formalized less than four months later through the unanimous endorsement of Superfund listing for the BPMD by local, county, and state officials.

ANIMAS RIVER STAKEHOLDERS GROUP

In the pre-spill context, the Animas River Stakeholders Group (ARSG) largely sustained the opposition to Superfund in the Upper watershed. In fact, the group's formation coincided with the mid-1990s surge in US Environmental Protection Agency (EPA)-led discussions over potential Superfund listing. The ARSG's official formation in 1994 represents the organizational expression of concerns of local residents, including mine operators and owners, about losing local control over implementation and enforcement of water quality standards in the watershed (Garrison 2015). The founding members—William Simon, Stephen Fearn, and Peter Butler—all had extensive knowledge and histories of involvement with water quality and mining issues in the Upper watershed ("Coordinators' Bios" 2017).

Simon completed a PhD in evolutionary ecology at the University of California Berkeley and later founded Animas Environmental Services, a company that provides a wide range of environmental engineering services—including site assessment, remediation, water sampling, risk assessment, and environmental permitting (Animas Environmental Services,

LLC 2021). Simon was elected a San Juan County commissioner from 1985 through 1989 and at the time of this writing served as ARSG's coordinator emeritus.

Stephen Fearn completed BS degrees in mechanical engineering and business at the University of Colorado Boulder and became a professional engineer. He had decades of experience in the design, construction, and maintenance of numerous mine-related water quality improvement projects in the Upper watershed. Fearn was also a former mine owner, including the GKM until it was acquired by the San Juan Corporation in 2005, when Fearn's properties were foreclosed. He was also a proponent of reviving mining in the Silverton area. In the past, Fearn had expressed concern that Superfund designation would result in excessive litigation and bureaucratic delay, stigmatize the Town of Silverton, discourage tourism, decrease property values, and ignore local input and concerns. Fearn served as ARSG's co-coordinator until his death in April 2018.

Peter Butler earned a PhD in natural resource policy from the University of Michigan and has worked for a number of state and local agencies, including the Colorado Water Quality Control Commission (CWQCC), the Southern Ute/State of Colorado Environmental Commission, and the La Plata County Water Advisory Commission. He also served as ARSG's co-coordinator until Fearn's death in 2018.

While not an expressly anti-Superfund forum today, ARSG'S founders and longtime members consistently advocated for avoiding official designation because of distrust of the EPA, the State of Colorado, and local environmental groups (Coughlin et al. 1999). Regarding the latter, in 1994, San Juan County commissioner Bill Redd asked ARSG's initial facilitator, Gary Broetzman, "Do the crazy environmentalists from that crazy town downstream [Durango] have to participate" (Coughlin et al. 1999, 26).

ACCOMPLISHMENTS OF THE ARSG

The ARSG's substantive work began in 1995, when it started investigating sources of heavy metals contamination and remediation strategies following the adoption of official water quality standards for the Upper watershed by the CWQCC. In 2001, the ARSG completed a Use Attainability Study (UAS), which recommended water use classifications and quality standards to the

CWQCC. The UAS recommendations were adopted by the CWQCC and ultimately led to permanent water quality standards in the Upper watershed ("ARSG History" 2017).

Since being established, the ARSG has been a primary subsystem component and forum for AMD discussion and action in the Upper watershed. A large part of the ARSG's work has been to demonstrate that it is possible (and preferable) for AMD to be remediated without overt interference from the federal government. As discussed in chapters 3 and 9, the ARSG has advocated for a Good Samaritan (Good Sam) approach to restoration, whereby mostly nongovernmental and local entities take the lead in physically addressing the sources of AMD through on-the-ground projects.

Far from providing a long-term, comprehensive solution, the ARSG's approach has yielded effective results. In particular, the ARSG has coordinated successful efforts to relocate tailings piles away from tributaries of the Animas, constructed holding and settling ponds to capture and treat AMD prior to it entering such streams, and injected hydrated lime into standing pools of polluted water to increase alkalinity and improve overall mine pool conditions ("Summary of Mine Reclamation Projects" 2015).

The ARSG has also inventoried more than 400 historic mine sites in the Upper watershed. Of these, 33 mine adits (or openings) and 34 mine waste sites were determined to be responsible for more than 90 percent of AMD. These sites were prioritized for remediation through a twenty-year plan; as of November 2015, 18 reclamation projects sponsored or cosponsored by the ARSG had been completed ("Summary of Mine Reclamation Projects" 2015).

In the aftermath of the 2016 Superfund designation of the BPMD, the future role of the ARSG in the Upper watershed is uncertain. When asked about this, Butler responded, "I think it's really just up in the air. We don't know at this point. It'll make it more challenging to do any more remediation projects for sure" (quoted in Romero 2016).

SAN JUAN CLEAN WATER COALITION

The San Juan Clean Water Coalition (SJCWC) was formed in early 2015, just months before the GKM event. Its primary goal has been to support site-specific Good Sam legislation for the Upper Animas watershed. Following the 2015 spill, Good Sam legislation remained its top priority, yet the SJCWC indicated its willingness to support the Superfund process to clean up the bulk of heavy metals.

The SJCWC specifically advocates for the construction of a permanent treatment facility for AMD in Cement Creek. As part of its post-spill plan, the SJCWC sees itself as a representative of and an advocate for protecting the region's recreation- and tourism-based economy, as well as an information conduit between community interests and the federal and state agencies involved in watershed restoration ("Our Plan for the Animas" 2018). In many respects, it appears that the SJCWC void created by ARSG's organizational dissolution in August 2019.

TROUT UNLIMITED

The national conservation group Trout Unlimited (TU) has been a vocal supporter of the SJCWC. Locally, Ty Churchwell—TU's San Juan Mountains coordinator—has been a main source of this support. He currently serves as SJCWC's coordinator and is a leading force in promoting Good Sam legislation. As part of its renewed push for Good Sam legislation in the post-spill watershed, Churchwell (2015) stated the following in a press release:

> Our tried and true pollution cleanup laws, the Clean Water Act and Comprehensive Environmental Response, Compensation, and Liability Act (better known as "CERCLA" or "Superfund"), place the burden of cleanup squarely on the owners of the property. Generally, this is an excellent policy for most forms of pollution. But in the West, where the parties responsible for developing most of the old mine sites are long-gone, cleaning up these sites is a legal quagmire. Trout Unlimited will be coordinating the on-the-ground effort to educate communities and stakeholders while gathering supporters.

ARSG Participants and Additional Coalition Supporters

In large part, ARSG supporters have included past and present hardrock mine operators and owners in the Upper watershed; many of them may be identified as *potentially responsible parties* (PRPs) following the Superfund designation. Among these is the Denver-based Sunnyside Gold Corporation (SGC). The SCG owned the Sunnyside Mine, which it bulkheaded at the American Tunnel between 1999 and 2002 in order to reduce AMD. This closed the mine and stopped the discharge of polluted water directly from its opening. As previously discussed, this action has long been suspected of causing other adjacent mines to either begin AMD or increase ongoing discharge from neighboring mines—including GKM. According to a regional nonprofit environmental research organization, it was "very possible" that a physical connection between the Sunnyside and Gold King Mines existed, which allowed AMD from the former to the latter, ultimately causing water accumulation and the blowout at GKM (Garrison 2015).

Because of this connection, the SGC is widely seen as a primary PRP in the BPMD—along with its parent company, Kinross Gold Corporation, the world's fifth largest gold producer in 2019. It was no coincidence that the SGC was the lead petitioner in an ultimately unsuccessful appeals case filed by the Mountain State Legal Foundation (MSLF) against the EPA and its endorsement of the BPMD listing.[2]

Claiming that the court's previous ruling in support of the BPMD invited "agency [EPA] abuse and overreach in listing [Superfund] decisions" (Westney 2018), the MSLF case had two main questions. First, was it legal for the EPA to aggregate the forty-eight mine sites into the BPMD site? Second, did the EPA aggregate the forty-eight mine sites to downplay the release it triggered at GKM and, in the process, usurp control over reclaimed mining sites previously managed by the State of Colorado and local governments (*Sunnyside Gold Corporation* 2017)? The DC Circuit Court of Appeals answered "yes" to the first and "no" to the second.

Another Superfund skeptic is Larry Perino, longtime spokesperson and reclamation manager for the SGC. Perino has criticized the EPA's post-spill approach as "increasingly heavy-handed and threatening" and "coercive," with "built-in inefficiencies, burdens, and costs" (quoted in Pendley 2017). Roughly five months after the GKM event, he commented: "Superfund is potentially 'fool's gold'—there are real questions as to whether Silverton would even be eligible to be listed as a Superfund site, and even if it was listed, it's far from certain that monies would be available under Superfund" (quoted in Olivarius-Mcallister 2018).

And then there is Todd Hennis, another EPA skeptic and longtime advocate of reestablishing mining in the Upper Animas Basin. As current owner of GKM and owner at the time of the 2015 spill, Hennis supports the scenario in which AMD from the sealed (i.e., bulkheaded) Sunnyside Mine migrated to neighboring mines through a maze of subterranean paths, faults, and fissures. EPA records support this connection, citing that until 2003, GKM was discharging on average 7 gallons of AMD per minute; after 2003, GKM's discharge increased to hundreds of gallons per minute (Environmental Protection Agency 2016).

When the EPA announced in 2014 that drainage from GKM had reached a point that a remediation project was necessary, Hennis claimed he was coerced by the federal government to allow the EPA and its contractors access to his property. He stated: "EPA decided it [remediation at GKM] was too big a job for that year, so they piled many, many tons of earth and rock on the [GKM] portal, to quote, prevent a blowout during the winter. In doing so, they created the blowout conditions this year [2015]. I was forced to sign this agreement, under threat of Federal Court action . . . EPA has a limitless legal budget, so there is effectively no way a private citizen . . . can effectively fight the seizure of one's private land" (quoted in Romero 2017).

SILVERTON RESIDENTS

Many longtime Silverton residents also expressed steadfast opposition to Superfund designation prior to the 2015 spill, motivated by a mix of distrust of the federal government and environmental groups. Even after the spill, local opposition was sustained by fears of reduced property values, loss of local control, and, for many residents, a genuine desire to see hardrock mining return to the region—a sentiment currently evidenced by the following proclamation on the Silverton Chamber of Commerce's website: "[M]ining in Silverton closed down in the 1990s. However, there's still gold and silver in those mountains and rumor has it that mining will be back one day" ("Silverton History" 2018).

In addition, post-spill comments from select Silverton residents to the EPA echoed deep skepticism regarding the federal agency and its operations. For example, one commenter stated: "The last thing Silverton needs is the EPA involved. Such a bloated organization obviously in this case has done more harm than good. After someone hurts you, you don't come back for more so they can hurt you again." Another commenter included this statement: "The NPL option leads into a very legalistic, cumbersome and slow process that frequently takes many years . . . [and] generally results in significantly higher costs and timeframes for cleanup" (Esper 2018a).

LOCAL BUSINESS LEADERS

Sustained opposition to Superfund listing by the local business community has come from Aaron Brill, founder of the area's largest employer—the Silverton Mountain ski area, founded in 2001. Prior to the spill Brill commented: "Who wants to take vacation in a Superfund site? The image of a Superfund site can't be understated. We're supportive of clean-water initiatives, but we're not supportive of the damage that can be done from the perception of a Superfund site" (quoted in Williams 2011).

However, when compared to other Colorado ski areas, Brill's fears appear unsubstantiated. In particular, Vail and Beaver Creek have thrived for years despite being located near a significant source of AMD in the nearby Eagle River.

Perhaps the timeliest criticism of the EPA and the Superfund process came from Dave Taylor, a former geologist and part-time prospector from New Mexico. Shortly after his retirement to Silverton, Taylor wrote a letter to the editor of the *Silverton Standard* in which he presciently warned of the 2015 GKM spill a week *prior* to its actual occurrence. Moreover, he accused the EPA's assistant Region 8 administrator, Martin Hestmark, of instigating the spill to impose Superfund listing on Silverton. In particular, the letter stated: "We [EPA] will have to build a treatment plant at a cost to taxpayers of $100 million to $500 million . . . I believe that has been EPA's plan all along. The proposed [mine] plugging plan has been their way of getting a foot in the door to justify their hidden agenda for construction of a treatment plant. After all, with a budget of $8.2 billion and 17,000 employees, the EPA needs new, big projects to feed the beast and justify their existence" (Taylor 2015).

ELECTED OFFICIALS AND GOVERNMENT AGENCIES

As discussed in chapters 3 and 9, in both pre- and post-spill political debates, most Colorado Republicans focus on passing Good Sam legislation, while Democrats generally support reform of the 1872 General Mining Law. This general pattern relates to Superfund listing as well.

For example, former Republican Congressman Scott Tipton—whose district (CO-3) includes both San Juan and La Plata counties, has cited loss of local control and negative economic impacts as reasons to oppose Superfund. As an alternative, Tipton has supported and introduced multiple Good Sam-based legislative proposals. In reference to his 2018 Good Sam bill, coauthored with US Senator Cory Gardner (R-CO), Tipton commented:

"There are many good Samaritan groups that have the technical expertise, financial ability and desire to conduct successful remediation at abandoned mines," Tipton said in a statement. "But they are discouraged from taking on projects due to current regulations" (quoted in Romero 2018).

Alternatively, in reference to legislation he had recently cosponsored (the Hardrock Mining and Reclamation Act of 2015), US senator Michael Bennet (2015) (D-CO) remarked: "Hardrock mining is a part of our heritage in Colorado, but it is long past time to reform our antiquated mining laws. This bill would provide the resources necessary to help clean up the thousands of abandoned mines in Colorado, improve water quality, and prevent a future disaster for downstream communities."

Outside of Colorado, elected Republican lawmakers as well as candidates for national office have expressed opposition to Superfund and furthermore have used the 2015 GKM spill as material to criticize the EPA. For example, two lawmakers at the state level in Utah alleged that federal environmental officials deliberately triggered the 2015 spill to force Superfund listing; they also petitioned the state's attorney general to investigate such a possibility (Maffly 2015). Less than two weeks after the GKM spill, Republican presidential candidate Dr. Ben Carson capitalized on the sensational event at a campaign stop in Durango. Carson, standing near the banks of the Animas River, touted his opposition to Superfund, suggesting that it might hurt the town's reputation. Regarding the EPA, Carson stated that the agency is dominated by "a bunch of bureaucrats who don't know a bunch of anything," adding that under a Carson administration "you would wouldn't have to sue the EPA, because I would get rid of all the old people and bring in people who understand the Constitution" (quoted in Olivarius-Mcallister 2015).

Also at the federal level, former President Donald Trump and many of his top administrators have been critical of or outright opposed to the EPA. Trump's early budget priorities included reducing EPA staffing and funding (e.g., elimination of 3,000 employees and budget reductions of $330 million—from $1.1 billion to $762 million annually).[3]

A Pro-Change/Pro-Superfund Coalition?

In the Lower watershed, there is currently strong support for Superfund and the impending federally led remediation of abandoned mine sites in the BPMD. The main actors in this coalition are local politicians, everyday citizens, environmental organizations, downriver Native American communities, and members of the recreation- and tourism-based community of Durango, Colorado.

Prior to the 2015 GKM spill, however, this pro-change coalition was far less organized, visible, and active when compared to the competing one led by the ARSG. This observation is based largely on the complete absence of an organized forum such as the ARSG. This paucity of organized support for Superfund in the Lower watershed was unexpected; prior to initiating research for this project, it was assumed that such a support base would have existed.

Notwithstanding, it is possible to posit a set of beliefs consistent with the post-spill policy preference of the pro-change, pro-Superfund perspective. To varying degrees, these include normative support for tenets of regulatory capitalism (government intervention), social liberalism (activist government), and traditional cooperative federalism (integrated programs and collaborative working relations between federal and state governments). In addition, multiple tribal nations are located in the Lower watershed; in the post-spill context, they have endorsed NPL listing and espoused additional preferences for self-determination and security, as well as defense for unique spiritual beliefs related to water and traditional ways of life. Concerns over environmental justice on behalf of Native American communities have grown more intense the farther down the river flows in the watershed (for detailed discussion, see chapter 8, this volume).

The fact that very little organized pre-spill support for Superfund existed in the Durango area and downriver Native American communities is significant. Indeed, a fundamental assumption of the ACF is that in most policy disputes, there exist identifiable coalitions organized around broad and competing policy preferences—and that these coalitions exist in some organized form prior to a compelling or dramatic focusing event. Even in the pre-spill context, the initial assumption was that there would have been two identifiable, cohesive coalitions organized around each main perspective on AMD remediation (i.e., pro-Good Samaritan/anti-Superfund versus pro-Superfund). Findings from this research suggest something different; rather than a well-developed policy subsystem with two clear coalitions—as suggested by the ACF—it is likely that a nascent or bifurcated subsystem existed in the pre-spill lower Animas River Basin, one without real congruence between beliefs and coordinated activity unique to each coalition.

This observation is bolstered by close inspection of the official definition of an *advocacy coalition* according to the ACF. The definition is strict and has two parts; coalitions must have an ideological component (i.e., a belief system) *and* a behavioral component (i.e., coordinated and organized activity over time) (Wieble et al. 2011, 352). While both components were identifiable in the Upper watershed's anti-Superfund coalition, they were far less concrete in the competing pro-Superfund coalition—that is, until the fateful day of August 5, 2015.

Pre-Spill Superfund Support

Perhaps the most outspoken critic of the ARSG and its generally anti-Superfund position both prior to and following the GKM spill is Travis Stills, a Durango-based environmental attorney. According to Stills, most proponents of Good Sam legislation have also been the main opponents of Superfund listing. This is supported by positions expressed by groups such as the ASRG, Trout Unlimited, and the Colorado Mining Association. Stills sees such groups as "either patsies of the likely PRPs or paid consultants to the PRPs"; they have operated primarily as "PRP-shielding effort[s]."[4]

Another longtime supporter of Superfund both before and after the GKM event is Jennifer Thurston, director of the Telluride, Colorado–based Information Network for Responsible Mining (INFORM). The nonprofit organization provides public information and education about unsafe and irresponsible mining practices ("About" 2017). While INFORM's work is focused on permitted mines and proposals, not inactive or abandoned mine reclamation, Thurston suggested that its work in the Animas Basin has increased since the Superfund designation. Similar to Stills, Thurston sees the PRPs, particularly the SGC and the lobbying efforts of the ARSG, as the primary forces of opposition regarding Superfund designation.[5]

A third identifiable source of Superfund support prior to the 2015 spill is 4Corners Riversports, a Durango-based rafting and retail store located on the banks of the Animas ("About Us" 2018b). Owner Andy Corra has held the position that given the size and cost involved with the AMD problem in the Upper Animas watershed, Superfund listing has always been the only option for comprehensive restoration.[6] In October 2015 testimony before the US Senate Committee on Small Business and Entrepreneurship, Corra stated: "I understand it makes a lot of people nervous. But the NPL is the only clear path forward. It can be done in a sensitive manner" (quoted in Esper 2018b).

To bolster this, Corra referenced the thriving tourist mecca of Moab, Utah, home of the US Department of Energy's (DOE) Moab Tailings Project Site, where 16 tons of uranium tailings are being removed from the banks of the Colorado River. Although not officially an NPL site, an estimated $720 million has been spent removing the 439-acre, 94-foot-tall pile of contaminated soils from an area that is clearly visible from highways leading both to and from Moab.

In Durango itself, a reclaimed Superfund site is located mere steps from the bustling downtown tourist district. Along the southwestern bank of the Animas River, the area known as the Durango Mill Site was first contaminated by a large smelter in which lead, silver, and gold ores from the Silverton area were processed from 1881 to 1920. The site was later used as a processing mill for uranium and vanadium and production of yellow cake from 1942 to 1963. The cumulative impact of these activities was serious contamination of land and water from heavy metals, radioactive materials (most notably radium), and toxic chemicals used in uranium processing (sulfuric acid, sodium chloride, and ammonium sulfate). By 1959, an estimated 490 million gallons per day of leaching solution, containing these and other chemicals used in the processing of uranium ore, were discharged into the Animas. Comprehensive cleanup under Superfund lasted from 1986 to 1991. No lasting stigma had significant negative effects on business development or real estate prices; on the contrary, Durango's overall economy blossomed following Superfund listing (US Department of Energy 2019).

Post-Spill Support for Superfund Listing

Due to the severe discoloration of the Animas from the 2015 spill and the intense and sustained media coverage that followed, a spike in support for

Superfund designation quickly developed, as numerous parties endorsed the decision to officially list the BPMD. The effect of the highly visible and dramatic focusing event remains undeniable.

DURANGO AND LA PLATA COUNTY ELECTED OFFICIALS

In January 2016, the La Plata County Board of Commissioners unanimously approved a resolution of support for Superfund designation in the Upper Animas watershed. This predated the vote of approval by their San Juan County counterparts. The Durango City Council also expressed unanimous support for Superfund listing in a January 2016 letter to then Colorado governor John Hickenlooper (D). In it, then mayor Dean Brookie (D) expressed concerns over water quality for Durango residents. Brookie stated: "What Durango needs might be different from what Silverton needs . . . This is not to upstage Silverton in any way, but the 20,000 people on our water system, compared with the repair needed on our water system, creates vulnerability for next [2017] summer . . . This is fairly urgent on our part, and independent of Silverton action" (quoted in Pace 2016).

SAN JUAN CITIZENS ALLIANCE (SJCA)

Formed in 1986, the SJCA is a leading Durango-based environmental group, focused on advocacy for clean air, pure water, and healthy lands ("About Us" 2018c). As recently as 2011, the SJCA was of the position that the EPA should withhold a Superfund designation while the collaborative community-based processes of the ASRG sought to address the issues of metals loading at the Animas headwaters.[7] According to the group's river keeper, Marcel Gaztambide, this stance was partially the product of organizational capacity; the SJCA lacked resources to advocate for Superfund designation in the years preceding the 2015 spill. Instead, the group relied on the ARSG to mitigate problems of AMD while its own work revolved mostly around wilderness protections in the San Juan Mountains, Wild and Scenic River status for the Animas, and other protections that would prohibit future mining activities in the region.[8] Notwithstanding, the SJCA has been in complete support of Superfund listing since the 2015 spill.

SILVERTON AND (OTHER) AREA RESIDENTS

In the Upper Animas watershed, organized support for Superfund designation prior to the spill was virtually nonexistent. Rather, most pre-spill support came from the Lower Animas watershed. Nonetheless, community support in favor of the listing emerged in an opinion piece printed in the *Silverton Standard* on August 20, 2015, roughly two weeks following the spill. The piece was enthusiastically endorsed by dozens of local residents ("Some Silvertonians OK with Superfund" 2015).

Less than two months prior to the listing of the BPMD on Superfund's NPL, the EPA had received a mere thirty-three public comments on the proposed site. Of those comments, 55 percent were supportive of Superfund, 15 percent opposed the listing, while 30 percent took no official stand on whether to list the BPMD. Silvertonian Michael Constantine, who owns property near the site, stated: "For decades now, I've listened to those born into mining culture pine for the return of large scale mining. Well, here it is, but in reverse—the same kind of work, same kind of pay; probably better benefits. I welcome both the economic stimulus and the change to one day fish in orange Cement Creek, which flows near both my mining claim and my house in Silverton, six miles downstream" (quoted in Esper 2018a).

NATIVE AMERICAN TRIBES

The Lower Animas River flows directly through the western third of the Southern Ute Native American Reservation. The Southern Ute Indian Tribe (SUIT) was notified by the State of Colorado on the day of the GKM spill and immediately dispatched its emergency response team and alerted the State of New Mexico. The toxic plume took approximately three days to reach the northern border of the SUIT Reservation.

According to Curtis Hartenstine, Water Quality Program (WQP) manager for SUIT, the tribe was not actively supporting Superfund designation prior to the 2015 event, primarily because metals pollution was not considered a significant issue since the reservation is located a considerable distance downriver from the Silverton area.[9] Hartenstine said SUIT had sporadically detected elevated concentrations of aluminum in water samples, yet they were considered to be minor and ambient since the metal is found in the majority of drainages on the reservation.[10]

According to the tribe's recent "Update on Animas River Health" ("Gold King Mine Spill" 2017), its WQP began coordinating with the EPA, the State of Colorado, La Plata County, and other local entities to direct sampling and monitoring locations prior to the plume's arrival on the reservation. Together with historical data (collected since 1992), these pre-spill measures of water quality and macroinvertebrate populations were compared with samples taken during the spill. Findings indicated increased metals concentrations and lower-than-average pH levels on the day the spill arrived that quickly fell to pre-spill levels. Subsequent monitoring revealed no residual impacts from the GKM spill ("Gold King Mine Spill" 2017).

The Animas joins the mainstem San Juan River near Farmington, New Mexico, after which it flows (from east to west) across the 27,000-acre Navajo Reservation, covering parts of northwestern New Mexico and southeastern Utah. The tribe did not officially endorse Superfund designation until June 3, 2016, almost a year after the GKM spill ("Navajo Nation" 2016).

Diné C.A.R.E.—Citizens against Ruining Our Environment—is an all-Navajo environmental organization based on the reservation. It was founded in 1988. The mission is to educate and advocate for the tribe's traditional teachings and way of life. Its main goal has been to empower the Navajo people to organize, advocate, and control their own destinies ("About Us" 2018a). While not engaged in direct advocacy for Superfund listing prior to the GKM spill, the group subsequently endorsed the BPMD Superfund listing. Prior to the spill, Diné C.A.R.E. was focused predominantly on regulating coal-fired power plants and nuclear waste sites, as well as its forest and groundwater resources ("About Us" 2018a).

Tó Bei Nihi Dziil (TBND) is a group of Navajo and other indigenous community organizers concerned primarily with water and food security; they focus mostly on empowering youth leaders. TBND was established prior to the GKM spill and, coincidentally, was facilitating a community teach-in during the 2015 spill. In the ensuing days, TBND shifted focus and quickly mobilized resources in anticipation of the spill's arrival on the Navajo Reservation.[11]

As discussed in chapter 8, Navajo farming communities near Farmington and Shiprock, New Mexico, were particularly affected by the 2015 GKM spill. In the aftermath, residents described negative impacts and associated concerns related to spirituality and ceremonial practice, food safety and security, tribal sovereignty, environmental justice, and poor communication in addition to more immediate concerns over water quality and the safety of irrigation water taken directly from the San Juan River. Given these myriad impacts to the Navajo Tribe and the reliance of many communities on irrigated agriculture, it was surprising to learn that little to no organized support existed for Superfund listing and comprehensive AMD remediation prior to the 2015 spill. This lack of proactive support for preemptive cleanup may be related to a number of factors, including a lack of information on the severity of AMD coming from the Upper watershed, poor communication among state or tribal jurisdictions, and a failure to coordinate comprehensive planning activities. It is likely that these factors were compounded by a spatial dimension, one related to the sheer distance (both physical and cultural or actual and perceived) between the arid reservation and the high-alpine reaches of the Upper Animas River watershed.

CONCLUSION

This chapter examined the two main policy preferences regarding AMD remediation and abandoned mine restoration in San Juan County, Colorado. One centered on continued opposition to federal Superfund designation in the Upper watershed and support for locally led, stakeholder-driven efforts by Good Samaritan parties such as the ASRG that had been the norm for decades. The other centered on securing Superfund designation to ensure comprehensive long-term abatement of AMD and restoration of abandon mines. The chapter also sought to identify and organize the primary beliefs and actors behind these competing perspectives into coalitions, in the tradition of the Advocacy Coalition Framework.

The expectation was that distinguishable coalitions existed in the Upper and Lower Animas River watersheds, both prior to and following the GKM incident and resultant toxic plume. Moreover, it was expected that in the aftermath of the highly visible and dramatic spill, the efforts of the pro-change coalition in support of Superfund would be significantly bolstered by the intense media and political attention—perhaps to such a level that the inertia supporting the status quo would be overcome. These expectations were partially confirmed, as this research revealed that an organized and active coalition generally in favor of preserving the status quo existed in the Upper watershed well before the 2015 spill, yet a comparable coalition supporting change did not exist in the Lower watershed prior to the mine blowout and severe discoloration of the Animas River.

In this post-spill context, the pro-status quo coalition articulated by the ARSG and others was able to briefly sustain its opposition to Superfund, yet the balance of power had shifted. In part, this stemmed from an emergent pro-change coalition led by elected officials, interest groups, tribal entities, and community activists in the Lower watershed. Buoyed by the powerful national and international reactions, as well as the widespread media coverage following the spill, calls for a strong, comprehensive, federally led response mounted. Significant policy change soon followed with the NPL listing of the BPMD, as many prominent leaders in the Upper watershed reversed their decades-long opposition to Superfund designation. Superfund was unanimously approved by both the Silverton Board of Trustees and the San Juan County commissioners. Yet in lieu of the August 5, 2015, GKM spill, none of these developments would likely have occurred, and never in such a rapid and absolute fashion.

NOTES

1. As a political scientist, I understand ideology as a broad belief system of ideals and ideas regarding the appropriate role and scope of government in society, which provides a basis for normative thoughts on political theory, economics, and public policy. Ideology is best viewed as a quadrant where specific beliefs are located, as opposed to a strictly connected, linear set of beliefs.

2. *Sunnyside Gold Corporation v. United States Environmental Protection Agency* was filed by the MSLF on April 26, 2017, in the US Court of Appeals, DC Circuit. On June 4, 2018, the DC Circuit ruled against the SGC, stating that it would not reconsider its prior decision to back the EPA's addition of the entire BPMD to the federal Superfund list.

3. Paradoxically, Donald Trump's selection for EPA administrator, Scott Pruitt, has expressed support for remediating and redeveloping the nation's 1,340-plus Superfund sites. In fact, prior to the announced reductions at the EPA, Pruitt pledged to urge the Trump administration to increase the budget for Superfund (through an infrastructure spending bill) and criticized the delayed status of projects that have been under remediation for decades. See Koss (2018).

4. Travis Stills, email to the author, March 26, 2018.

5. Jennifer Thurston, email to the author, March 23, 2018.
6. Andy Corra, email to the author, January 2, 2018.
7. Marcel Gaztambide, email to the author, December 13, 2017.
8. Marcel Gaztambide, email to the author, December 13, 2017.
9. Curtis Hartenstine, email to the author, December 22, 2017.
10. Curtis Hartenstine, email to the author, December 22, 2017.
11. Dr. Becky Clausen, email to the author, March 21, 2018.

REFERENCES

"About." 2017. Information Network for Responsible Mining. http://www.informcolorado.org.

"About Us." 2018a. Diné C.A.R.E. https://www.dine-care.org/about_us.

"About Us." 2018b. 4CornersRiverSports. https://www.riversports.com.

"About Us." 2018c. San Juan Citizens Alliance. http://www.sanjuancitizens.org/about-us#whatwedo.

Animas Environmental Services, LLC. 2021. "Services." https://animasenvironmental.com/services/.

"ARSG History." 2017. Animas River Stakeholders Group. http://animasriverstakeholdersgroup.org/blog/index.php/history/.

"Bennet Introduces Bill to Reform Antiquated Hardrock Mining Laws." 2017. Michael Bennet US Senator for Colorado, September 19. https://www.bennet.senate.gov/public/index.cfm/2017/9/bennet-introduces-bill-to-reform-antiquated-hardrock-mining-laws.

Churchwell, Ty. 2015. "A Renewed Push for Good Samaritan Legislation—Colorado Trout Unlimited." August 24. http://coloradotu.org/2015/06/a-renewed-push-for-good-samaritan-legislation/.

"Coordinators' Bios." 2017. Animas River Stakeholders Group. http://animasriverstakeholdersgroup.org/blog/index.php/coordinators-bios/.

Coughlin, Christine W., Merrick L. Hoben, Dirk W. Manskopf, and Shannon W. Quesada. 1999. "A Systematic Assessment of Collaborative Resource Management Partnerships." Master's project, University of Michigan, Ann Arbor.

Environmental Protection Agency. 2016. *One Year after the Gold King Mine Incident: A Retrospective of EPA's Efforts to Restore and Protect Impacted Communities.* https://www.epa.gov/sites/production/files/2016-08/documents/mstanislausgkm1yrreportwhole8-1-16.pdf.

Esper, Mark. 2018a. "Most Comments Favor Superfund Listing." *Silverton Standard,* May 15. http://www.silverstonstandard.com/news.php?id=902.

Esper, Mark. 2018b. "Will Gold King Spill Taint Town's Economy?" *Silverton Standard,* May 23. http://www.silvertonstandard.com/news.php?id=860.

Garrison, Steve. 2015. "Gold King Mine Spill a Disaster Waiting to Happen." *Albuquerque Journal*, August 16. https://www.abqjournal.com/628976/gold-king-mine-spill-a-disaster-waiting-to-happen.html.

"Gold King Mine Spill: 2017 Update on Animas River Health." 2017. Southern Ute Indian Tribe. https://www.southernute-nsn.gov/wp-content/uploads/2017/06/20170630-GoldKingMineSpill-Update.pdf.

Jenkins-Smith, Hank, Daniel Nohrstedt, Christopher Weible, and Karin Ingold. 2018. "The Advocacy Coalition Framework: An Overview of the Research Program." In *Theories of the Policy Process*, 4th ed., edited by Christopher Weible and Paul Sabatier, 134–148. New York: Westview.

Koss, Geoff. 2018. "Pruitt Wants Money for Water Projects, Superfund Cleanups." *E&E Daily*, May 22. https://www.eenews.net/eedaily/stories/1060050723.

Maffly, Brian. 2015. "Utah Lawmakers Speculate Feds Might've Orchestrated Toxic River Spill, Ask AG to Investigate." *Salt Lake Tribune*, August 8. https://www.sltrib.com/news/2015/08/19/utah-lawmakers-speculate-feds-mightve-orchestrated-toxic-river-spill-ask-ag-to-investigate/.

"Mission." 2017. Animas River Stakeholders Group. http://animasriverstakeholdersgroup.org/blog/index.php/goals-objectives/.

"Navajo Nation Endorses Superfund Cleanup of Colorado Mines." 2016. CBS4 Denver, July 11. http://denver.cbslocal.com/2016/07/11/navajo-nation-superfund-cleanup-colorado-mines/.

Olivarius-Mcallister, Chase. 2015. "Ben Carson Draws a Big Crowd in Durango." *Durango Herald*, August 8. https://durangoherald.com/articles/1477.

Olivarius-Mcallister, Chase. 2018. "Is Silverton Ready for a Cleanup?" *The Journal* (Cortez, CO), March 21. https://www.the-journal.com/articles/23931.

"Our Plan for the Animas." 2018. San Juan Clean Water Coalition. http://www.sanjuancleanwater.org/animas/.

Pace, Jessica. 2016. "Durango Sends Letter to Colorado Governor in Support of Superfund." *Durango Herald*, February 3. https://durangoherald.com/articles/2016.

Pendley, William Perry. 2017. "Gold King: EPA's Two-Year Rolling Disaster and a Path Forward to Fix It." Mountain States Legal Fund. https://www.mountainstateslegal.org/docs/default-source/default-document-library/gold-king-and-superfund-final.pdf?sfvrsn=bbedoac2_0.

Romero, Jonathan. 2016. "Mine Cleanup Organization Could Dissolve with Silverton Superfund Approval." *Durango Herald*, February 24. https://durangoherald.com/articles/2034-future-of-animas-river-stakeholders-group-uncertain.

Romero, Jonathan. 2017. "Though His Name's in Ink, Todd Hennis Said His Signature Appears Not by Choice." *The Journal* (Cortez, CO), March 6. https://the-journal.com/articles/28438.

Romero, Jonathan. 2018. "Gardner, Tipton Introduce Mine-Cleanup Bill." *Durango Herald*, December 6. https://durangoherald.com/articles/253671.

Sabatier, Paul. 1988. "An Advocacy Coalition Model of Policy Change and the Role of Policy Oriented Learning Therein." *Policy Sciences* 21 (Fall): 129–168.

"Silverton History." 2018. Silverton, Colorado, Chamber of Commerce. http://silvertoncolorado.com/index.php?page=page_41.

"Some Silvertonians OK with Superfund." 2015. *Silverton Standard*, August 20. http://www.silvertonstandard.com/system/news/pdf/standard8-20-2015.pdf.

"Summary of Mine Reclamation Projects in the Upper Animas River Basin." 2015. Animas River Stakeholders Group. http://animasriverstakeholdersgroup.org/blog/index.php/2015/11/04/387/.

Taylor, Dave. 2015. Letter to the Editor. Silverton Standard, June 23.

US Department of Energy. 2019. *Durango, Colorado, Processing and Disposal Sites Fact Sheet*. Washington, DC: US Department of Energy.

Westney, Adam. 2018. "Mining Co. Asks DC Circ. to Revisit EPA Superfund Ruling." *Law360*, July 27. https://www.law360.com/articles/1035431.

Wieble, Christopher, Paul Sabatier, Hank Jenkins-Smith, Daniel Nohrstedt, Adam Henry, and Peter deLeon. 2011. "A Quarter Century of the Advocacy Coalition Framework: An Introduction to the Special Issue." *Policy Studies Journal* 39 (3): 349–360.

Williams, David. 2011. "Thar's Gold in Them Thar Silverton Hills; Lead, Zinc, and a Lot Less Trout Below." *Colorado Independent* (Denver), August 20. http://www.coloradoindependent.com/97927/thars-gold-in-them-thar-silverton-hills-lead-zinc-and-a-lot-less-trout-in-rivers-below.

AFTERWORD

We All Live Downstream

From Gold Medal to Gold Metal *Waters, Lessons from the Gold King Mine Spill*

ANDREW GULLIFORD

There are many lessons to learn from the August 5, 2015, Gold King Mine spill, which unleashed 3 million gallons of toxic waste including arsenic, lead, copper, cadmium, and aluminum cascading down Cement Creek toward the Animas River, across three states, and into Lake Powell. Let's start with the first lesson—how to become a millionaire.

I don't mean the dollars that may flow to liability lawyers who lined up like stocked trout leaping at Power Bait. I mean the original nineteenth-century capitalists, those who "tamed" the Old West and brought railroad lines and small towns into Colorado's high country. A few made fortunes off the 1872 Mining Law, which has yet to be amended or repealed. Anyone can go forth and make money on this country's public lands. Across the West, the history of the nineteenth century and a large part of the twentieth century was to make money from extractive industries—grazing, lumbering, and, of course, mining. So here's the first lesson: privatize your profits and socialize your losses. That's how to become rich.

DOI: 10.5876/9781646421756.c011

In other words, take the money and run. Let someone else, generations hence, clean up the mess. It's worked for the gold, silver, lead, and copper companies from the nineteenth and twentieth centuries. Certainly, the uranium mines in the Four Corners region left a polluted legacy during and after World War II. Let's hope in decades to come we're not worrying about underground poison plumes from twenty-first-century fracking for deep pockets of natural gas. Oh well, let someone else worry about that.

As I stood on the bank of Cement Creek in Silverton the day of the blowout, I watched orange waters swirl and wondered what the impacts would be. Certainly, this is a black eye for the US Environmental Protection Agency (EPA), whose contractors had no emergency plan in place and no way to contact local first responders.

Not the sheriff. Not the Colorado State Patrol. No one knew what was hurtling downstream. How's that for disaster planning?

An hour by highway from Silverton, Durango boasts gold medal fishing waters as the Animas River winds through town. Tourists can fish almost out of their hotel windows. Following the insect hatches, dry fly fishermen enter the river wearing hip-high waders. Fisherfolk don long-sleeved lightweight shirts; with their caps and poised poles, they are as ubiquitous as Great Blue herons. A regional office for the national conservation organization Trout Unlimited has a Durango headquarters. The local chapter, named Five Rivers, remains devoted to conserving cold-water fish and fisheries. And then came the orange sludge roiling down the Animas like a scene from a B-grade Hollywood horror film.

Former Congressman Scott Tipton asked if the EPA would give itself a fine. The irony is inescapable. The flow continues out of mines near Gladstone close to Silverton at 300 million gallons annually and at 500 gallons per minute from the Gold King's #7 portal. The current stop-gap measure includes holding ponds to settle heavy metals—ponds right in the middle of winter avalanche paths.

The San Juan County commissioners do not have a cleanup master plan and for years avoided Superfund status. Now, damage has been done. What that will mean for the county's tourism and property values is unknown. Locals tell me that "folks don't realize down below" how complicated this is. I can only imagine. Who will define the liability of "potentially responsible parties"?

County sentiment has changed. After years of denying Superfund status for a major environmental problem, San Juan County now seeks adequate funding for a serious cleanup. The glaring spotlight of national and even international publicity within forty-eight hours of the spill, with over 1,000 media requests, has moved on. Getting the US Congress to fork up millions will not be easy because there's no pot of gold at the end of the polluted plume.

The single photo that galvanized the world was of three kayakers, including one of my former students, floating in an orange river. That image by *Durango Herald* photographer Jerry McBride was published all over the world. Rivers are supposed to be blue or green or maybe brown with silt, but never neon orange.

How ironic that the kayakers were paddling for eco-recreation when the very water under their boats turned toxic. And where were they photographed? Near Baker's Bridge, named after the leader of the first expedition into the San Juan Mountains in search of precious minerals. Silverton would be plotted and laid out in Baker's Park where Cement Creek and Mineral Creek join the Animas. Eventually, the Gold King Mine would be discovered because the Baker Expedition had shown the potential for gold and silver ore in the San Juans. More than a century and a half later, Cement Creek would vomit heavy metals in a distinctive orange toxic broth.

Talk about a case study in unintended consequences. The EPA contractors at the Gold King site—with no security, no satellite phone, and no emergency protocols—were checking a pile of dirt, a berm, they'd erected the year before. Then all hell broke loose. Or it seems that way. Even the Navajo Nation is suing the EPA because Native American farmers and ranchers fear diverting water for agricultural purposes. "They are not going to get away with this," Navajo Nation president Russell Begaye told tribal members at an emergency meeting in the Shiprock Chapter House. The Navajo Nation asked for $160 million in damages. They received a little over $1 million, but the lawyers are not finished.

Political grandstanding included visits by former Governor John Hickenlooper weeks later, who took an empty glass and drank from the Animas River. The attorneys general of Colorado and Utah did an inspection tour, as did tribal leaders.

"As the plume moves on, we're seeing a downward trajectory toward pre-event conditions," stated the EPA director, who admitted that the EPA accepts full responsibility. In Durango, the Animas is green again, but long-term effects on fish and other aquatic life continue to be studied.

Because of the river's steep gradient, the real damage will occur many miles downstream in New Mexico and Utah, where contaminants will eventually settle. It's a tri-state eco-disaster. How fitting that some refer to the Animas's Spanish name—the Rio Animas de las Perdidas, or "river of lost souls."

The Old West of no rules, no regulations, and every man for himself has left a legacy of 160,000 abandoned mines, 33,000 of which leak acidic toxins. Estimates are that 40 percent of Western headwaters may be contaminated. The price tag for cleanup—a hefty $35 billion. Where did we go wrong?

....

An answer to that question is to go back to our historical understanding of water in the West. That knowledge begins with a one-armed major. John Wesley Powell lost an arm as a Union captain at the Battle of Shiloh in the Civil War. As a veteran, he began to explore the American West, first by climbing Long's Peak north of Denver and then by launching an expedition down the Green River to the Colorado and all the way through the Grand Canyon in the wrong wooden boats. He would have done better with almost any other kind of boat design, but he chose a Great Lakes boat with a keel. Bad move. Too many rocks in swift rivers.

But he learned. His men rowed and portaged all the way through the Grand Canyon in 1869. He returned in 1871 and partially replicated his epic journey, all the while taking notes and understanding water in the West. Later, as director of the US Geological Survey, Powell would try to slow down the homestead rush for free land to make sure there was enough water for irrigated small-scale farming. He had radical visions for the West that butted head-on with politicians.

The major wanted Western political boundaries to reflect watersheds. He did not want square states and square counties; he wanted legal lines to follow the contours of watersheds and flowing water. Hence, the Gold King Mine spill, in Powell's plan, would have occurred not in La Plata County, with its boundaries, but rather in a different county that embraced the entire watershed of the Animas, Piedra, Pine, Navajo, Mancos, and La Plata Rivers—which all flow into the San Juan. He wanted local control of water. I believe local control would at some point have resulted in local awareness of polluted water and the potential for pollution solutions.

Powell didn't get his way. He resigned because of harassment by politicians championed by Nevada senator "Big" Bill Stewart, who resented Powell's interference with wild, unchecked Western settlement. Stewart sought rampant growth. If settlers failed because of a lack of water for agriculture, so be it. Stewart championed mines, miners, and the 1872 Mining Act, which brought about the Gold King Mine in the first place. Under that law, the Gold King went from an exploratory vein on public domain to private property.

Colorado water law is "first in time, first in use," which is the doctrine of prior appropriation. But Colorado water flows into creeks and streams and rivers that enter other states. The Animas River flows into the San Juan River, a major tributary of the Colorado that travels 383 miles and becomes one of the muddiest rivers in North America—once carrying 25 million annual tons of sediment and draining 25,000 square miles.

Over half of the watershed courses through tribal lands, specifically the 17-million-acre Navajo Nation, which has water rights on 600,000 acre-feet. No wonder the tribe became upset over an orange polluted river plume flowing into the San Juan, or their "Old Age River"; "Old Man River," or "Male Water" (Tooh Bika/ii); "Male with a Crooked Body" (Bika^ii Bitsiis Nanooltl'iizhii), or "One with a Long Body" (Bits'ns Nineezi). In Navajo, the San Juan has many names.

Congress failed to heed Powell's warnings in his seminal "Report on the Lands of the Arid Region of the United States" (1879). Finally, in 1922, with the Colorado River Compact, the Western states divided the entire Colorado River system into an Upper basin—which includes Colorado, Wyoming, Utah, and New Mexico—and the Lower basin of Arizona, California, and Nevada. The compact is all about water rights, and it set allocations in millions of acre-feet, with no references to pollution traveling downstream. Now, almost a century later, perhaps one of the lessons from the Gold King Mine spill into the Animas River would be to re-look at that river compact and finally address both industrial mining and agricultural runoff creating water pollution.

Under current laws like the Wild and Scenic Rivers Act (1968), the National Environmental Policy Act (1970), the Clean Water Act (1972), and the Endangered Species Act (1973), such a reappraisal is long overdue and will require not only political will going forward and tribes at the table but also an understanding of the past. For the Gold King Mine spill, a historical assessment and eco-volume already exists.

....

We learned the hard way that there is municipal water, agricultural water, recreational water, and tribal water and it is all the same. The Old West and the New West have collided. Damage from the Gold King Mine spill eventually resulted in a September 2016 Superfund listing by the Environmental Protection Agency of the Bonita Peak Mining District of thirty-five mines (including the Gold King), seven tunnels, four tailings ponds, and two study areas. After three decades of discussion, the citizens of Silverton and the residents of San Juan County, Colorado, knew they had to do something, and only federal resources would help. One local told me he was "almost glad it happened" because it exposed the need for change.

Navajos using the Animas River as it merged with the San Juan River in New Mexico remained leery. The University of New Mexico College of Pharmacy landed a $3.5 million grant to study exposure of Native communities to metal mixtures from mines. Urgency for the grant came because of the Gold King spill. A year later, Native farmers remained cautious about letting river water irrigate their thirsty crops.

The US Supreme Court weighed in on the State of New Mexico's lawsuit against Colorado. Utah waited in the wings to see what the legal liability was of 880,000 pounds of metals that churned downstream and turned the river bright orange-yellow due to mercury, nickel, zinc, copper, lead, cadmium, and arsenic. The Animas enters the San Juan in New Mexico. Then the San Juan flows across southeast Utah, with a discharge volume of 2.5 million acre-feet—twice the volume of the Rio Grande—before it is trapped in the placid waters of Lake Powell.

"Science needs to communicate results more clearly and, as a community, we need to become more comfortable as science consumers," argued Fort Lewis College biology professor Shere Byrd. "As citizen-scientists, we should be asking the scientists how they are interpreting the data, what information that interpretation is based on and why they interpreted the results in this way," she stated in a *Durango Herald* editorial (Byrd 2016).

The Obama administration's Environmental Protection Agency vowed to make things right. Damages would be compensated. A cleanup would occur. But on January 13, 2017, under the Trump administration, the EPA said it was not liable for the seventy-three claims totaling $1.2 billion it had received. As for the contentious but finally agreed upon Superfund cleanup, who knows? Back to lesson #1 in Western history. Make money off the land. Leave. Ignore any industrial or corporate damages. Instead, spend your dollars on Washington, DC, lobbyists.

....

The Gold King Mine blowout polluted the Animas River through the town of Durango and deleted summer tourist dollars. Three years later, in the summer of 2018, in the midst of searing, extreme drought, the 416 Fire did the same thing and thousands of visitors stayed away. In 2015, local residents learned about the impacts of water pollution. In 2018, they learned bitter lessons about air pollution as constant smoke settled in the valleys and slipped downstream, following the Animas all the way to Aztec, New Mexico.

How dry? So dry that the USGS water gauge at Aztec measured only 5 cubic feet per second. The river had become barely a line of wet stones. A blowout similar to the 2015 disaster would have had even more ominous consequences because the water wouldn't have been there to carry away heavy metals. They would have settled much higher upstream.

The Gold King Mine blowout was not the first on the Animas, of course. In June 1978, miners tunneling under Lake Emma high above Silverton accidentally came too close to the bottom of the lake. The roof of the stope of the Sunnyside Mine collapsed on a Sunday night, draining the lake through the mine and on into the Animas River. On a blessed Sunday, with no one working, the lake burst through—dumping thousands of gallons of water into the mine, flooding it, and blowing mine debris, sludge, heavy metals, gray oozing mud, and rough-cut mine timbers all the way down to Farmington, New Mexico. Silverton miners credited their salvation to a large 12-ton, 16-foot-tall Italian marble statue, *Christ of the Mines*, which they had erected in 1959.

No one paid much attention then to a high-elevation mine blowing pollution into the Animas. It had been an occasional occurrence. Now we know better.

At sunrise shortly after the Gold King spill, Ute leader and elder Kenny Frost led a prayer vigil at the river's edge, quietly intoning "Water is sacred. Water is life." He slowly waved eagle feathers. The US National Park Service sent out a press release.

"Contaminated wastewater released from a mine on the Animas River in Colorado is expected to arrive at Lake Powell via the San Juan River sometime early next week," stated Christiana Admiral on August 7, 2015, for the Glen Canyon National Recreation Area. She continued: "While containment levels and impacts anticipated at Lake Powell are not yet understood, as a precautionary measure, the National Park Service at Glen Canyon National Recreation Area is encouraging visitors to avoid drinking, swimming, or recreating on the San Juan river within Glen Canyon National Recreation Area and on the San Juan arm of Lake Powell until further notice."

Until further notice. When, exactly, will we notice and correct the eco-errors of the past? Of the many lessons to be learned from the Gold King spill, and they will continue, was the slogan I saw three weeks after the spill while canoeing on the Green River below Moab, Utah. Because of the West's outstanding scenery, a couple had brought their canoe all the way from St. Louis. A sticker on the stern simply said, "We all live downstream."

REFERENCES

Admiral, Christiana. 2015. "News Release: Glen Canyon Officials Urge Caution in San Juan Arm of the Park Following Release of Contaminants from an Upstream Colorado Mine." *Grand Canyon River Guides* (blog), August 7. https://www.facebook.com/permalink.php?story_fbid=1123228267692620&id=256500817698707.

Byrd, Shere. 2016. "Gold King Mine Spill: How We Can All Become Better Citizen-Scientists." *Durango Herald*, August 6. https://durangoherald.com/articles/107973.

Powell, John Wesley. 1879. "Report on the Lands of the Arid Region of the United States, with a More Detailed Account of the Lands of Utah, with Maps." 2nd ed. Washington, DC: Government Printing Office. https://pubs.usgs.gov/unnumbered/70039240/report.pdf.

Contributors

ALANE BROWN, professor emeritus, Department of Psychology, Fort Lewis College

BRIAN L. BURKE, professor, Department of Psychology, Fort Lewis College

KARLETTA CHIEF, assistant professor and assistant specialist, Department of Soil, Water, and Environmental Sciences, University of Arizona

STEVEN CHISCHILLY, graduate, Environmental Studies, Fort Lewis College

BRAD T. CLARK, associate professor, Department of Political Science, Fort Lewis College

BECKY CLAUSEN, associate professor, Department of Sociology and Human Services, Fort Lewis College

MICHAEL A. DICHIO, assistant professor, Department of Political Science, University of Utah

BETTY CARTER DORR, professor, Department of Psychology, Fort Lewis College

CYNTHIA E. DOTT, professor, Department of Biology, Fort Lewis College

GARY L. GIANNINY, professor, Department of Geosciences, Fort Lewis College

DAVID A. GONZALES, professor, Department of Geosciences, Fort Lewis College

ANDREW GULLIFORD, professor, Department of History, Fort Lewis College

LISA MARIE JACOBS, ALERT (A Locally Empowered Response Team), Berkeley, California

PETE McCORMICK, professor, Environmental Studies, Fort Lewis College

ASHLEY MERCHANT, MSW, Graduate School of Social Work, University of Denver

TERESA MONTOYA, assistant professor, Department of Anthropology, University of Chicago

SCOTT W. ROBERTS, aquatic biologist, Mountain Studies Institute, Durango, Colorado

LORRAINE L. TAYLOR, assistant professor of management, School of Business Administration, Fort Lewis College

JACK TURNER, Animas Community Listening and Empowerment Project

KEITH D. WINCHESTER, graduate, Tourism and Hospitality Management, Fort Lewis College

MEGAN C. WRONA, assistant professor, Department of Psychology, Fort Lewis College

JANENE YAZZIE, Sixth World Solutions, Lupton, Arizona

Index